# Contents

# Foreword

Public perceptions of smoking and nonsmoking have changed drastically. Ten years ago, although less than half the population smoked, smoking was still considered the norm and nonsmokers who objected were considered social deviants. Today the situation has reversed. Most people—including most smokers—regard cigarette smoking as the air pollution that it is.

As a result, efforts to reduce or eliminate smoking in public places have been accelerating. In some parts of the country, nonsmoker protection is mandated by law. In California alone, approximately ten million workers are covered by laws protecting them from the toxic chemicals in secondhand cigarette smoke. In other places, employers have voluntarily restricted smoking to improve working conditions and, incidentally, to increase productivity and profits.

There is, in fact, a strong consensus that nonsmokers should be protected. A recent Harris poll revealed that 80 percent of the adult population (including 75 percent of the smokers polled) favored legal restrictions on where one could smoke.

Given this consensus, why must many workers still breathe other people's smoke? The main reason is that most people don't realize the consensus exists. In particular, many managers worry that nonsmoker protection will cause disharmony and decrease productivity. (Of course the tobacco industry does what it can to foster this mistaken impression.)

*The Smoke-free Workplace* addresses this concern and many others in a comprehensive way. Using real examples from around the country, it demonstrates that employees welcome smoking restrictions and regard them as improvements in the work environment. It illustrates how easy it is to develop and implement these policies. And it shows how smoking restrictions *save* money for employers. In short, this book makes it clear that the most difficult step toward providing a smoke-free workplace is developing the will to act.

As employers come to know this, the pace of progress will accelerate even faster.

Stanton A. Glantz, Ph.D.
President, Californians for
Nonsmokers' Rights
Associate Professor of Medicine,
University of California at
San Francisco

# Editor's Note

As a psychiatrist, I work mostly with individual patients in a small office. During the early 1970s, when smoking was permitted in my consultation room, I began to realize that cigarette smoke would sometimes irritate my sinuses. However, there were two reasons why I hesitated to stop patients from smoking in my presence. First, people tend to talk more freely when they are at ease, and cigarettes seemed to help some patients relax. Second—and perhaps more important—I did not wish to seem dictatorial.

After several years of internal debate, I finally decided that my own health and comfort were more important. At first, the ban was limited to my consultation room, but today I do not allow smoking in my waiting room either. My sinuses no longer get irritated, my wife no longer complains that I smell of smoke, and—because cigarette ashes no longer get strewn about—my office needs cleaning less than half as often.

Why did I take so long to protect myself? Why do people often hesitate to ask smokers to be more considerate? The answer is that advertising has conditioned our society to the belief that smoking is customary and tolerable. This strategy has been so effective that even nonsmokers who despise tobacco smoke may allow smoking in their homes.

Public opinion is shifting, however, as more and more people realize that drifting smoke is something they can well do without. If someone were to squirt you with a water pistol every time you rode a bus or dined in a restaurant, few would question your right to take action to halt this annoyance. I believe that non-smokers have just as much right to stop the annoyance of breathing smoke-filled air. Increasingly, they are banding together to get state and local governments to restrict or prohibit public smoking. More than two-thirds of our states have enacted laws to restrict smoking in public areas. Many cities have clean indoor air laws, and many businesses have voluntarily instituted non-smoking rules.

In her book *A Smoking Gun: How the Tobacco Companies Get Away with Murder*, Dr. Elizabeth M. Whelan suggests a parallel with public spitting. During the late nineteenth century, it was quite common for men to expectorate tobacco and phlegm onto the nearest available surface. Then it was discovered that the health of "nonspitters" could be harmed by germs present in the spit. "Do Not Spit" signs were posted, the National Tuberculosis Association (now called the American Lung Association) organized a war against public spitting, and laws were passed prohibiting it.

No doubt many spitters objected strenuously to this encroachment on their freedom, but spitting soon went from being a socially acceptable custom to one which drew general public contempt and disgust. Nowadays, people seldom spit in public, and one would have to look long and hard to find a spittoon. Perhaps smoking and ashtrays will also become rare in the not too distant future. *The Smoke-free Workplace* is your step-by-step guide to bringing this day closer.

Stephen Barrett, M.D.

# About the Authors

William L. Weis, Ph.D., C.P.A., is associate professor and chairperson of the Department of Accounting, Albers School of Business, Seattle University, and is a health and management consultant with Rosner, Weis and Lowenburg, Inc. He has written on all aspects of workplace smoking control and has helped many prominent corporations develop and implement their policies toward smoking.

Bruce W. Miller, who lives in Seattle, Washington, consults and writes about the smoke-free trend in our society. Since 1980, his *Nonsmoker's Assertiveness Guide* has encouraged nonsmokers to speak up for clean air. He also writes for a variety of national magazines on business and health topics. He developed and managed the first statewide birth defects surveillance program and has been a consultant to federally funded programs in the fields of education and criminal justice.

# About the Editor

Stephen Barrett, M.D., who practices psychiatry in Allentown, Pennsylvania, is a nationally renowned consumer advocate, author, and medical editor. He edits *Nutrition Forum Newsletter* and has produced more than 20 books on health topics for the general public. In 1984 he received the FDA Commissioner's Special Citation Award for Public Service.

# Acknowledgments

We appreciate the many individuals who shared with us their experiences, resources, and enthusiasm for clean air. Three experts deserve special thanks for reviewing the manuscript and making helpful suggestions: Regina Carlson, executive director of New Jersey Group Against Smoking Pollution (GASP); Alan Blum, M.D., editor of the *New York State Journal of Medicine* and co-founder of DOC (Doctors Ought to Care); and Stanton A. Glantz, Ph.D., president of Californians for Nonsmoker's Rights. And long-time friend Bob Potts was a great source of encouragement.

# 1

# The Awakening Battleground

By 1911, when the mass-produced cigarette was still fairly new, many nonsmokers were aware that "comfort" for smokers meant discomfort for themselves. That year Dr. Charles G. Peace organized the Non-smokers' Protective League, whose purposes he explained in the November 10th issue of the *New York Times*:

> The league does not seek to abridge the personal rights of anyone, but it does seek to awaken the sense of fairness in those who use tobacco and to impress upon them the fact that they have not the right to inflict discomfort and harm upon others.
>
> That tobacco smoke and the odor of tobacco are irritating to normal, unpoisoned respiratory membranes is attested to by the personal experience of thousands of persons who are daily, and in many cases hourly, forced to inhale them. They produce headache, dizziness, nausea, and even fainting; they injure the eyes and lungs, the nervous and alimentary systems, and in other ways they cause harm, discomfort, and pain.

Peace's description still rings true today. But instead of quietly suffering, nonsmokers are saying *"No"* to smoky environments—in public and at work. Nonsmokers' rights groups are multiplying and have been winning victory after victory to clear the air. City ordinances are being passed to prohibit smoking in public places. And restaurants are being pressed to provide clean air as well as clean food.

The most intense struggle is taking place in the smokers' last stronghold: the workplace. Here smokers and nonsmokers must confront each other because they have no choice. Unlike patrons of a restaurant, most workers cannot simply walk out.

That smoking takes its toll on the health of smokers has been obvious for many years. One merely needs to observe a smoker cough or wheeze through the day to see the evidence. But smoking has been such an accepted part of American life that, until recently, few people paid attention to how it hurts nonsmokers.

Nonsmokers today *are* paying attention. We rub our burning eyes—and know that work would be more pleasurable and productive without the smoke. We see cigarette burns in the carpets. We ponder the black ring around the ventilation register that delivers the air we breathe. We wonder whether the salesman who coughs throughout his sales pitch will be the one to service the account. And, when traveling by air, we wonder whether a cigarette butt dropped in the wrong place will spark another airline disaster.

Even with these thoughts, the idea that workplaces should be entirely smoke-free may be uncomfortable to both employees and management—not because they like smoke, but because they fear change. Rather than expend energy solving this problem, many defend or excuse the status quo. Nonsmokers may say, "It's really not that bad." Smokers declare that "smoking gives me something to do during my break." And management decides, "Banning smoking would upset too many people."

Actually, people who think this way are fooling themselves. Do they really believe that keeping unhealthy practices is better than promoting healthier ones? Or that an organization will lose by promoting the health of its employees?

The norm *should* be a smoke-free work environment. But, if applicants ask whether smoking is allowed on the job, some employers perceive them as troublemakers. Supporting this view are smokers who claim that banning smoking would infringe on their personal liberty and who turn a deaf ear to the argument that their smoking imposes air pollution on others.

Frank Wetzel, executive editor of the Bellevue (Wash.) *Jour-*

*nal American,* responded wisely to a reporter's concerns that liberties were on the way out when smoking was banned in the newsroom:

> Your comment about the next freedom to go tells me that you have missed the whole point. We aren't dealing with something that "bothers" people, for heaven's sake; we are dealing with something that is harmful to their health.

Whether management likes it or not, the move for nonsmokers' rights is irreversible. The "right" to smoke on the job is no longer unquestioned. What will management do with this new challenge? Ignore the issue and foster resentment by nonsmokers? Fire nonsmokers for their justified objections? Play musical chairs with smokers and nonsmokers? Wait for smokers to quit or die, and hope that no more apply for work? Hope that clean air advocates will find other jobs? Or take the lead in resolving an issue before it gets too hot and interferes with work more than the smoke already does?

The nonsmokers' rights movement is not only raising a moral issue but also calling attention to the high costs of permitting smoking on the job—and of having smokers on the payroll, even if they don't smoke at work.

In this book you'll find compelling statistics that show how portions of an organization's profits can go up in smoke. We hope that, when you finish it, you'll think differently about your smoky workplace. If you're a manager, we hope you will question whether hiring smokers is good for business. If you're a smoker, we hope you will understand why getting a new job may become increasingly difficult. If you're a nonsmoker, we hope you will be inspired to press for more healthy working conditions. No matter what your position, you'll find ample information that justifies clearing the air.

Some owners of companies that are now smoke-free knew

simply from their sense of moral correctness that prohibiting smoking on the job was the right thing to do. The health of their employees was sufficient justification. Health is still a good reason. This book will back up that reason with the economic one: smoking on the job costs money.

In a tangle of emotional, psychological, and managerial problems, a big question mark should light up the ledger books. How much is smoke costing the organization? It is an interesting question, and one that more and more organizational leaders are answering for themselves: *Too much!*

# 2

# Why Smoke-free Workplaces?

"Our policy at Radar Electric was quite simple," explained Warren McPherson, president of Radar Electric in Seattle, Washington, from 1972 to 1983. "We hired only nonsmokers and prohibited smoking on company premises—and that applied to visitors and customers as well as employees."

Why break the time-honored tradition of allowing, and sometimes even encouraging, smoking on the job? For any or all of these reasons:

- to reduce absenteeism
- to increase productivity
- to improve morale
- to improve the health and well-being of employees
- to improve compliance with OSHA's mandate to provide a safe work environment
- to protect the company's investment in human resources
- to protect furniture and equipment
- to reduce maintenance and cleaning costs
- to reduce insurance costs
- to reduce the risk of industrial accidents and fires
- to reduce the cost of ventilation
- to enhance organizational image
- to improve the appearance of the organization's buildings and grounds
- to reduce employee turnover
- to reduce the risk of offending customers.

"Many of these were important fringe benefits of our policy," said McPherson, "but I actually instituted the policy for personal reasons. You see, my mother had lung cancer, and I wanted to fight back at the enemy that tortured and finally killed her. This

19

policy was my way of doing that. I just wanted my company to be smoke-free."

So add one more reason to be smoke-free:

● to generate personal satisfaction from confronting what Surgeon General Everett Koop has proclaimed "the most important public health issue of our time."

Clearly, the effects of smoking are costing organizations a bundle. Those that already forbid smoking on work premises have reported decreased expenditures for cleaning, maintenance, and replacement of furniture, carpeting, and equipment. So it's not surprising that more and more employers are thinking twice about allowing smoking on the job and about hiring smokers.

Reinforcing this trend is the nonsmokers' rights movement, which has increasingly turned its attention toward the workplace. Recent research supports what nonsmokers have known all along: that in addition to being irritating and annoying, smoke can damage the health of nonsmokers. The number of complaints, grievances, worker's compensation claims, and lawsuits filed by nonsmokers is on the increase. Indeed, the number one job-related complaint in the 1980s is a smoky work environment.

## Managers Understand

Many managers sympathize with the nonsmokers' plight and are distressed that smoking costs their organization money. A survey in the Seattle area in 1981, for example, found that 88.6 percent of the managers wanted smoking banned or segregated (see Table 2:1).

Table 2:1. Workplace Policy Preferred by Managers

| | |
|---|---|
| No restrictions | 42 (11.3%) |
| Separate areas for smokers and nonsmokers | 202 (54.4%) |
| Total smoking ban on premises | 127 (34.2%) |

Source: 1981 survey by W. L. Weis and C. P. Fleenor.

Of the managers with hiring responsibility, 53.4 percent preferred to hire nonsmokers even before being informed of the greater absenteeism rates of smokers. When the managers were asked to assume that absenteeism was 50 percent greater among smokers—a small cost to employers compared with others discussed in this book—82.5 percent of the managers said they would favor nonsmoking job applicants. And when the health hazard of smoke to nonsmokers was added to the assumptions, 88.8 percent of the managers found nonsmokers more appealing (see Table 2:2).

Table 2:2. Preferences of Managers Toward
Similarly Qualified Job Applicants

| Assumptions | Prefer Smokers | Prefer Nonsmokers | Undecided |
|---|---|---|---|
| None | 1 (0.4%) | 119 (53.4%) | 103 (46.2%) |
| Smoking increases absenteeism | 1 (0.4%) | 185 (82.5%) | 37 (16.6%) |
| Smoking increases absenteeism and is a health hazard | 1 (0.4%) | 198 (88.8%) | 23 (10.3%) |

Source: 1981 survey by W. L. Weis and C. P. Fleenor.

The twenty top executives, people who could immediately order development of a policy to restrict smoking on the job, all said they would hire a nonsmoker instead of a smoker because of the greater absenteeism rates of smokers and the health hazards of drifting smoke. Of the twenty, eighteen (90%) favored smoking restrictions, eight (40%) favored a total ban, and ten (50%) wanted distinct and separate smoking areas.

## Why Managers Hesitate

With these preferences among managers with power, why aren't smoke-free workplaces more common? Many of the managers in the survey revealed in informal discussion that they thought smoking at the workplace is a right that would be illegal to prohibit. Some thought that a policy of not hiring smokers would violate equal employment opportunity laws. And a few questioned whether smoking at work really costs an employer money.

This book answers these questions and more. It makes clear the costs of smoking at the workplace and provides the basics of planning and implementing a smoke-free work policy. After reading it, you'll know how to refute these myths:

- Smokers will quit their jobs.
- Businesses will lose customers.
- Smokers usually retaliate against employers who ban smoking.
- It is illegal to hire only nonsmokers.
- Banning smoking on the job is illegal.
- A no-smoking policy will reduce profits.
- Employers should fear smokers more than nonsmokers.
- Smokers do not want smoking restrictions on the job.
- No-smoking policies destroy employee morale.
- Employers prefer unrestricted smoking in the workplace.
- Secondhand smoke is not a health hazard to nonsmokers.
- Employees are not legally entitled to a healthy, smoke-free work environment.
- Employee morale is higher where smoking is permitted.
- Labor unions oppose smoking restrictions on the job.
- Smoking bans are impossible where employees are organized by strong unions.
- Smoking on the job is a right.
- Nonsmokers don't care if they work in smoke.
- Management has nothing to fear by ignoring nonsmokers' requests for smoke-free work environments and grievances against smoke in the work environment.

# 3

# Smoking and Absenteeism

Despite personnel literature replete with articles on the causes, consequences, and cures of absenteeism, personnel theorists rarely seem to consider that employees who take sick leave might simply be sick. Instead, absenteeism is attributed to reasons ranging from management style to the newest and trendiest "stress" syndromes invented by state-of-the-art psychologists.

Whatever the reasons people stay away from work, one simple fact is clear: Smokers do so substantially more often than nonsmokers. Nonsmokers who have observed their co-workers for many years are not surprised by this fact. But many employers who have not would be startled by the differences in absenteeism rates shown in the tables below. Table 3:1, based on data in the 1981 *Statistical Abstract of the United States*, compares the number of days lost per year by those who have never smoked and smokers with various levels of cigarette consumption. Table 3:2, based on data in *Smoking and Health: A Report of the U.S. Surgeon General's Office* (1979), compares the number of workdays lost by smokers and nonsmokers in various age groups. And Table 3:3 summarizes studies done on three groups of employees.

Note that in the organizational studies, the differences between smokers and nonsmokers are greater than those in Tables 3:1 and 3:2. The data in *Statistical Abstracts* come from the Bureau of the Census, which gets its information from the National Health Interview Survey (U.S. National Center for Health Statistics), which relies on self-reporting. The data for 1979, reported in the *1981 Statistical Abstract*, were based on interviews with 157,461 persons. Persons may underreport their own

Table 3:1. Workdays Lost Per Year Related to Number of Cigarettes Smoked (Based on data in *Statistical Abstract of the United States*, 1981, p. 123)

| Extent of Daily Cigarette Use | Men | | Women | |
|---|---|---|---|---|
| | Days Absent | Increase over Nonsmokers | Days Absent | Increase over Nonsmokers |
| Never smoked | 4.9 | - | 6.1 | - |
| Under 15 | 8.5 | +73.5% | 8.2 | +34.4% |
| 15 to 24 | 9.6 | +95.9% | 7.7 | +26.2% |
| 25 to 34 | 6.8 | +38.8% | 9.3 | +52.5% |
| 35 or more | 7.0 | +42.9% | 12.3 | +101.6% |
| All smokers* | 8.5 | +73.5% | 8.2 | +34.4% |

*Includes those for whom number of cigarettes smoked is unknown.

Table 3:2. Workdays Lost Per Year by Smokers and Nonsmokers in Various Age Groups (Based on 1974 data tabulated in the 1979 Surgeon General's Report)

| Age | Never Smoked | Now Smoke | Smokers' Increase | |
|---|---|---|---|---|
| Men 17+ | 3.4 | 5.1 | 1.7 | +50.0% |
| 17-44 | 3.0 | 5.5 | 2.5 | +83.3% |
| 45-64 | 4.4 | 4.5 | 0.1 | + 2.3% |
| Women 17+ | 4.5 | 5.6 | 1.1 | +24.4% |
| 17-44 | 4.3 | 5.3 | 1.0 | +23.3% |
| 45-64 | 5.4 | 6.5 | 1.1 | +20.4% |

Table 3:3. Studies by Three Organizations

| Employer | Results |
| --- | --- |
| Dow Chemical Company[a] | Cigarette smokers were absent 5.5 more days per year than nonsmokers |
| Calif. Dept. of Water Resources[b] | Smokers took 50% more sick leave than nonsmokers |
| Major U.S. airline[c] | Smokers took 64% more sick leave than nonsmokers |

a. Three-year study of 1,352 employees at Dow's Midland Division.
b. Reported in *DWR News*, Sacramento, Calif., Feb. 1977.
c. Data from 6-month study in the airline's regional administrative offices sent anonymously to Dr. Weis. The average number of sick leave hours was 37.4 hours for nonsmokers and 61.2 for smokers. The study's supervisor confirmed the figures when telephoned, but said that the sensitivity of the issue prevented him from public discussion of the subject. The study had been undertaken in an attempt to prove that absenteeism rates for smokers were *not* higher than those for nonsmokers!

absenteeism rates, especially if they are sensitive to an implied causal relationship with a habit they have been unable or unwilling to break. The absenteeism rates for the airline and Dow Chemical studies came directly from personnel records, which are more reliable.

At Boyd Coffee, in Portland, Oregon, where smoking is banned everywhere except in the parking lot during regular work breaks, the company's 110 employees were absent a total of 175 days in 1980—an average of 1.6 days per employee. At the Austad Company in Sioux Falls, South Dakota, where smoking has always been prohibited, the company's 151 employees missed a total of 260 days during 1982—an average of 1.7 days per employee. These figures are strikingly lower than those in the National Health Interview Survey, which found that the average smoker missed 8.35 days in 1979. This rate was 5.3 times the rate for Boyd Coffee employees and 4.9 times that for Austad employees.

These figures even suggest that "passive smoking" (discussed in Chapter 7) increases absenteeism among nonsmoking employees—by impairing their health, their morale (by requiring them to work around smokers), or both.

## Cost of Sick Leave

What do these absentee rates mean in dollars and cents? In 1980, Dow Chemical calculated that excess wages were costing $657,000 per year for 2,804 smoking employees at an hourly wage rate of $5.23 per hour. Assuming there are 230 8-hour workdays per year, the excess wage cost per smoker would be $234 per year, excluding benefits and payroll taxes.

If a company's average payroll cost per employee is $30,000 (including benefits and taxes) for an average work year of 230 days (assuming 8 holidays, 3 weeks of vacation, and 7 days of sick leave, the average for smokers and nonsmokers combined), the average daily cost per employee, smoker or nonsmoker, is $130. Using this figure, smokers would cost their company the extra amounts shown in Table 3:4.

## The Real Costs Are Even Higher!

Absenteeism costs associated with smoking can be substantially greater when overtime is involved. In the steel industry, for example, the average daily cost per male steelworker is $200 at straight time. Absenteeism changes the estimated costs per year

Table 3:4. Additional Annual Payroll Costs of Smokers (Based on Data from 1981 *Statistical Abstracts of the United States*)

| Cigarettes Used Daily | Days Absent Attributable to Smoking | Extra Annual Cost Attributable to Smoking | |
|---|---|---|---|
| | | @ $130/day | @ $200/day |
| Males | | | |
| Under 15 | 3.6 | $468 | 720 |
| 15 to 24 | 4.7 | 611 | 940 |
| 25 to 34 | 1.9 | 247 | 380 |
| 35 or more | 2.1 | 273 | 420 |
| All males* | 3.6 | 468 | 720 |
| Females | | | |
| Under 15 | 1.5 | 195 | 300 |
| 15 to 24 | 1.6 | 208 | 320 |
| 25 to 35 | 3.2 | 416 | 640 |
| 35 or more | 6.2 | 806 | 1,240 |
| All females* | 2.1 | 273 | 420 |

*Includes those for whom number of cigarettes smoked is unknown.

by the amounts listed in the $200/day column in Table 3:4. But if the absentee's replacement is paid time and a half, the expense would be even greater.

Those assumptions are actually too conservative because they don't consider that employees usually make money for their companies. With professional services, for example, the billing rate can be several times the direct cost of the employee's services. A salaried CPA who is paid $20 per hour may have a client billing rate of $70 per hour—hence loss of that person's services during a peak work season could reduce the firm's profits by an additional $400 per 8-hour day.

The tobacco industry has argued (through a paid consultant) that sick-leave policies are designed so that absences do not represent an incremental cost to the company. (An example of this argument appears in "The Other Side of the Smoking Controversy," by L. Solomon, in *Personnel Administrator*, March 1983.) This viewpoint is nonsense. Greater rates of absenteeism *always* cost more than lower rates, even if a policy allows unlimited accumulation of leave, or at the extreme, compensates employees for not taking sick leave. The numbers above represent the cost of having more personnel than is necessary—the cost of being overstaffed to make up for excess work-loss days taken by smokers on the payroll.

Employee morale should also be considered when evaluating absenteeism that favors a minority at the expense of their co-workers. Most nonsmokers know that their smoking colleagues take sick leave more often and that the workload increases among the nonsmokers when the smokers are gone. Revulsion to this obviously unfair system led Warren McPherson at Radar Electric to replace his standard sick-leave policy with an extended vacation plan in which each employee is given so many days of paid time-off, regardless of the reason, whenever the employee chooses. His rationale:

> That way all of my employees could enjoy the additional time off from work each year. The smokers could spend the time in bed; the nonsmokers on the ski slopes. And now no one ever has to feign illness just because, for whatever reason, he wants or needs to take a day away from work.

Presumably the ski industry has benefited along with the employees, because Radar Electric hired its last smoker in 1977, and its workforce of 100 had only three smokers at the end of 1983.

Smoking's toll on health is enormous. But despite this well-documented fact and the logical connection between impaired health and absenteeism, many employers seem surprised when introduced to the absenteeism differentials between smokers and nonsmokers cited in this chapter. We therefore encourage employers to look at their own data. Those who compare a random selection of 100 employees who smoke and 100 who don't will find a difference in sick leave usage that may well convince them to implement a strict clean-indoor-air policy.

# 4

# Smoking and Productivity

"Not only did her smoking ritual shut down her own work—it shut down half of the office," complained Dale Stephens, vice president for data processing with a major West Coast bank. "Every time she struck a match, it was a signal that the time had arrived for another informal—and unauthorized—work break. Four or five of my staff members would move, in unconscious reflex, toward her desk to begin another 10-minutes of chitchat. Every single cigarette she lit was costing me an hour of lost productivity."

Not any more. Corporate headquarters told Stephens to establish whatever smoking policy he deemed appropriate for his department. So he announced a clear policy that smoking would no longer be tolerated during working hours.

Seattle restaurateur Robin Woodward discovered that the drywall workman she hired was spending 30 minutes of each billing hour on his smoking habit. "The painters weren't much better," she said. "They managed to stay on the ladders only 15 minutes at most between smoking breaks. Now my rule for subcontractors is firm: absolutely, positively no smokers."

Lyndon Sanders, owner of the Nonsmokers' Inn in Dallas and the Dollar Inn in Albuquerque, now hires only nonsmokers. His statistics show that they clean rooms 25 percent faster than smokers do.

## How Smokers Waste Time

Smoking rituals are often so masterfully executed that most people accept them as a part of the smoker's job description. And let's not forget the time lost looking for a cigarette, which can even take the smoker out of the workplace to buy another pack. Cleaning and fiddling with a pipe is also good for a major break at least twice during the day.

The tobacco industry argues that all workers waste time whether they smoke or not. Even if this is true, it wouldn't justify time lost due to smoking. The following story—told by a Seattle bank manager who wishes to remain anonymous—illustrates how smoking's role in wasting time is often overlooked at the worksite:

> For years I watched my smoking employees ceremoniously rise from their desks, light up, and gaze contemplatively out the window at Elliott Bay, or engage co-workers in diversionary gab, or just wander aimlessly among the desks as if preoccupied by some complex problem that demanded concentration impossible to achieve while sitting at a desk without a cigarette from which to draw strength and inspiration.
>
> And this all looked perfectly normal to me. Then one day my secretary inquired . . . whether I would fire non-smokers if they also "made the rounds" and "kept watch for enemy ships on Elliott Bay."
>
> Five minutes later, I watched a programmer as he made the rounds, and I tried to imagine him without his cigarette. I could barely control myself. Such dereliction may go unnoticed in smokers, but not in nonsmokers.
>
> I gave my secretary an immediate raise, announced that smoking during working hours would be banned as of the first of the following month, and—this is why I don't want you to use my name—decided there and then never to hire another smoker.

A smoking ban is also an effective way to keep break times equal among smokers and nonsmokers—especially when worker morale and productivity are important. Mark Miller is a maintenance man for Lincoln General Hospital in Lincoln, Nebraska.

He has reported that his boss criticized him severely for taking a few extra minutes for his break. "The boss said more or less that he wanted me to hurry up and get something done. It's all right for smokers to sit down or stop what they are doing and go have a smoke. Just because I don't smoke I'm getting penalized for taking a little longer on break." Miller says his morale went downhill and that his response is to "do what I have to do and that's it. I don't go out of my way to do anything extra."

David Hesketh, manager of Yonny Yonson's, a sandwich and frozen yogurt shop in Seattle, Washington, agrees that smokers take not only extra time but also extra breaks. When he was a cook at a Denny's restaurant, Hesketh says, "other cooks who were smokers were taking up to two smoking breaks an hour—which meant at least ten minutes of lost time that I somehow had to compensate for with faster work." Adding smoking waiters and waitresses to the picture, he concludes:

> The nonsmokers were substantially subsidizing their smoking colleague's down-time with virtually no work breaks and speed-ups . . . The inherent injustice in that environment was a continuous source of morale problems for those employees who were contributing the most to the restaurant.

At Yonny Yonson's, Hesketh hires "only heads-up people . . . bright and quick to serve . . . who, by and large, don't smoke," because "employee morale is very high and I intend to keep it that way. One smoker taking unscheduled work breaks—hence putting more pressure on my other employees—could destroy the morale we have worked hard to cultivate."

## Statistical Studies

How much time is actually lost to the smoking ritual? Several businesses and consulting firms have measured the amount of time the average smoker loses to the smoking rituals of lighting, puffing, staring, appearing deep in thought, and enjoying an informal workbreak. The Major Pool Equipment Company in Clifton, New Jersey, found that smokers lost 2 to 10 percent of their efficiency, depending on how frequently they smoked.[1] The average smoker wasted about 30 minutes a day to the smoking ritual—some 6 percent of the work year.

The Robert E. Nolan Company, a consulting firm in Simsbury, Connecticut, watched cigarette smokers waste 30 minutes per day, but found that pipe smokers wasted 55 minutes per day, or 11 percent of the work year.[2]

If we accept the finding that smokers waste 30 minutes a day, workers paid $25,000 per year would waste $1,500 of their employers' money, and 94 nonsmokers could do the same amount of work as 100 smokers. These figures do not include lost productivity of offended nonsmokers or their increased absenteeism (discussed in Chapter 3). Nor do they include billing time lost when astute clients protest.

While the smoking employee is gazing off into space, the client is still being billed at the usual rate. Occasionally a client will have the opportunity to observe this subtle form of theft and balk at payment, as did the production editor of a business journal. She timed how long the paste-up artist in the typesetting shop took to light and smoke cigarette after cigarette. Ten minutes of each hour went up in smoke. When she demanded a 6 percent reduction of the artist's bill, she got it. (She should have asked for 16 percent!)

When time lost to smoking is combined with time lost to absenteeism, the average smoker loses 17 days per year—which means that the work done by six smokers could be done by five nonsmokers.

# References

1. Carlson, R., *Smoking or Health in New Jersey: A Progress Report on Making Nonsmoking the Norm*, New Jersey State Department of Health, Occupational Health Program, 1981, p. 24.

2. *Wall Street Journal*, November 7, 1978.

# 5

# Lower Insurance Rates

Should nonsmokers and smoke-free organizations pay less for health, life, disability, fire, auto, and industrial accident insurance? America's insurance companies understand this. Their data show conclusively that nonsmokers pay more than their share for the smoking habits of others. This amounts to grand larceny!

Ponder these statistics:

• According to health economist Dr. Marvin Kristein, smokers burden the nation's health care system (particularly hospitals) *at least* one and a half times as much as nonsmokers.[1]

• Two other studies cited by Kristein found that industrial accident rates for smokers are twice that of nonsmokers.[2]

• According to a study by State Mutual Life Assurance Co. of America, death rates for smoking policyholders in all age categories are two and a half times more than for nonsmoking policyholders.[3]

• Figures from State Mutual also show that smokers are killed in fatal auto accidents at 2.6 times the rate of nonsmokers.[3]

• Thirty percent of building fires and 97.4 percent of fatalities in structural fires are attributable to careless smoking.[4,5]

## Cost of Health Insurance

How do these differences affect insurance premiums? In 1982, Group Health Cooperative of Puget Sound offered group coverage of hospital and outpatient care and prescription medication to employees for a monthly average of $55 per adult, smoker or nonsmoker. If an organization's benefit package covered spouses as well as employees, the average annual cost was $1,320 per employee.

Assuming that one-third of Group Health's adult membership smoked (about the same percentage of smokers as the national adult population in 1982) and that smokers used the health care system 50 percent more than nonsmokers did (as noted in item 1 above), the cost of insuring each smoker would have been $71 compared with $47 per nonsmoker. At these rates, an employer buying group health insurance as a fringe benefit would have paid an extra $24 per month per unmarried smoker ($288/year) or $48 per month ($576/year) for each smoking employee married to a smoker. Because cigarette smoke raises the incidence of respiratory disease in small children, the cost of insuring children in smoking households is also greater.

In another study, Luce and Schweitzer determined the difference between insuring smokers and nonsmokers by considering the dollar costs and percentages of cancers, circulatory diseases, and respiratory diseases caused by smoking.[6] According to their figures, smoking increased the annual health-care cost, adjusted to May 1982 dollars, by approximately $14.3 billion for 52.4 million smokers—$274 per smoker.

Because smokers are riskier to insure, many insurance companies offer nonsmoker discounts. In 1981, a report done by Regina Carlson for the New Jersey Department of Health noted that 32 companies offered discounts up to 20 percent on health, life, or auto insurance.[7] Today that figure is higher. An organization that prohibits smoking on company premises and hires only nonsmokers is obviously in a good position to bargain for discounts that reflect the differences in risk between smokers and nonsmokers and between smoking and nonsmoking environments.

In a landmark decision, the National Association of Insurance Commissioners adopted a resolution at its 1984 annual meeting calling for different health insurance rates for smokers and nonsmokers.

Large companies should also consider self-insuring to reduce the amount paid to outside insurers for health coverage. Under a partial program, a company obtains outside coverage to protect itself from disastrously large claims. By self-insuring for the first $15,000 of medical care for each employee, the Austad Company, which forbids smoking on company premises, spends one-third of what industry experts say it would pay for comparable commercial coverage!

## Loss of Experienced Workers

For most companies, the extra cost of insuring smokers is small compared with the cost of losing them to disability-related retirement or premature death. Mortality and disability tables show that smokers die more frequently than nonsmokers during years of peak productivity. The 1979 Surgeon General's Report compares mortality rates for male cigarette smokers and nonsmokers.[8] The data in Table 5:1 show, for example, that the smokers of two packs a day or more were 2.76 times as likely as nonsmokers to die between the ages of 45 and 54. The rates are even greater when smokers are compared with those who have never smoked—as is done in Table 5:3.

Table 5:1. Comparison of Death Rates of Smokers and Nonsmokers at Various Ages, Based on Data from the 1979 Surgeon General's Report (Nonsmoker baseline rate in each category is 1.)

| Cigarettes Smoked | Age 35-44 | Age 45-54 | Age 55-64 |
|---|---|---|---|
| 20 to 30 daily | 1.91 to 1 | 2.41 to 1 | 2.05 to 1 |
| 40 or more daily | 2.59 to 1 | 2.76 to 1 | 2.26 to 1 |

According to the same data, fewer than 60 percent of 22-year-olds who continue to smoke will live to age 60. But more than 85 percent of 22-year-old nonsmokers can be expected to live that long.

## Cost of Life Insurance

Cigarette smoking is the leading cause of preventable illness and death in the United States. According to a recent analysis by Dr. R. T. Ravenholt, director of World Health Surveys, Inc., 485,000 Americans died in 1980 of cigarette-related causes (see Table 5:2). Today he estimates that number to be over 500,000—more than one-fourth of all American deaths from all causes! Table 5:3 provides additional details on the risks of smoking.

Table 5:2. U.S. Deaths Attributable to Cigarettes (1980)

| | |
|---|---:|
| Cancer | 147,000 |
| Heart and blood vessel diseases | 240,000 |
| Lung diseases | 61,000 |
| Digestive system diseases | 14,000 |
| Infant mortality | 4,000 |
| Fires and other accidents | 4,000 |
| Other | 15,000 |
| Total | 485,000 |

Source: Ravenholt, R. T., "Addiction Mortality in the United States, 1980: Tobacco, Alcohol and Other Substances," *Population and Development Review* 10:697-724, 1984.

The State Mutual Life Assurance Company of America began offering discounted insurance to nonsmokers after the 1964 Surgeon General's Report showed that they lived longer than smokers. Since it first offered policies for nonsmokers, State Mutual has been comparing the death rates of its smoking and

Table 5:3. The Risks of Smoking.

| | |
|---|---|
| Overall risks | Smokers lose about five minutes of life expectancy for each cigarette they smoke. Smoking two packs a day decreases life expectancy more than eight years. Smoking one pack a day decreases life expectancy six years. |
| Heart and blood vessel disease | Nicotine in smoke raises the heart rate and contributes to high blood pressure. The incidence of heart attacks and strokes is increased. Women who smoke and use birth control pills are more than 10 times as likely to suffer a heart attack and 20 times as likely to suffer a stroke by brain hemorrhage as women who do neither. The incidence and severity of circulatory diseases of the arms and legs are increased. |
| Cancer | Lung cancer, the leading cause of cancer death, is more than ten times as common in smokers than in nonsmokers. The incidence of cancers of the larynx, mouth, throat, esophagus, bladder, pancreas, and kidney is increased. |
| Respiratory problems | Smoking is the leading cause of chronic bronchitis and emphysema. Lung function is measurably impaired in smokers. |
| Other problems | Smoking is a significant factor in causing and aggravating peptic ulcers. Maternal smoking during pregnancy significantly lowers birth weights and increases the risk of spontaneous abortion, stillbirths, and death during early infancy. Some 2,000 deaths and 4,000 injuries each year are due to cigarette-induced fires. |

Source: Whelan, E. M., *A Smoking Gun—How the Tobacco Industry Gets Away with Murder*, George F. Stickley Co., Philadelphia, 1984.

nonsmoking policyholders. In October 1979, the company published a study covering 1973 to 1978. During this six-year period,
the mortality rate of policyholders who smoked at the time the
policy was issued was 2.5 times the rate of nonsmoking policyholders. The difference was greatest for the youngest age groups,
as Table 5:4 shows.

Table 5:4. Comparison of Actual and Expected Death Rates of
Smokers and Nonsmokers (Based on payments made by State
Mutual Life Assurance Company of America. Expected rate in each
age group is 100.)

| Age at Issue | Smokers | Nonsmokers | Ratio |
| --- | --- | --- | --- |
| 20 to 29 | 180 | 26 | 6.9 to 1 |
| 30 to 39 | 205 | 29 | 7.1 to 1 |
| 40 to 49 | 122 | 48 | 2.5 to 1 |
| 50 to 59 | 92 | 77 | 1.2 to 1 |
| 60 or over | 105 | 60 | 1.8 to 1 |
| Average | 132 | 53 | 2.5 to 1 |

Keep in mind that because the study considered only policies
issued since 1964, a person between the ages of 20 and 29 at the
time the policy was issued would have to have died by age 43 *at
the oldest* to be counted in the mortality statistics. The authors of
the study attribute the decrease in smoker mortality ratios in the
older groups to cessation of smoking by many of those designated as smokers.

What did these differences mean in actual payments to State
Mutual's policyholders? For the 20-29 age group, the company
had $506 million at risk for its insurance policies on smokers, and
paid out $831,000 in death benefits on smokers, or .164 percent
of the total amount insured. Nonsmoker policies put the company
at a $874 million potential risk, but paid out only $195,000 on

nonsmokers, or .022 percent of the total amount insured. Thus for every $1.00 in death benefits paid on nonsmokers' policies from 1973 through 1978, $7.45 was paid on smokers' policies.

If we include policyholders who became insured between ages 20 and 39, the average payout rate for smokers was .325 percent of the total insured value; and for nonsmokers, it was .045 percent of the total. The payout rate for smokers was thus 7.22 times that for nonsmokers.

The study's authors emphasize that their results clearly justify discounted life insurance premiums for nonsmokers. After demonstrating that the average difference in life expectancy between smokers and nonsmokers is 7.3 years, they point out that this is significantly greater than the difference between men and women that has traditionally been the basis for the differences in premium rates between the sexes. Cowell and Hirst suggest that the increased life expectancy of women may reflect the fact that fewer women smoke.[2]

Two other points made in State Mutual's report bear mention. One is that the smoker-to-nonsmoker mortality ratios for its policyholders were much greater than those reported in the 1979 Surgeon General's Report. This discrepancy can be attributed to the greater death rate among the general population compared with the insured population from causes not related to smoking. Because the employee population resembles the insured population more closely than the general population, State Mutual's data are probably more important than the Surgeon General's for determining how smoking should affect life insurance premiums.

The second point the authors emphasize is the greater mortality ratios of many causes of death not commonly associated with smoking: mental disorders and diseases of the central nervous system (ratio 2.4 to 1), digestive diseases (5.8 to 1), suicide (9 to 1), and homicide (2.2 to 1). The authors suggested that riskier lifestyles might account for these mortality rates. When the causes of death not usually associated with smoking were eliminated, the authors found that the mortality ratio for smokers *increased* from 2.5 to 2.8.

As shown in Table 5:5, the figures are even more striking if former smokers are considered separately from those who have never smoked.

Table 5:5. Estimated Mortality Ratios of Current Smokers vs. Those Who Have Never Smoked

| Age When Policy Issued | Smokers | Never Smoked | Ratio |
|---|---|---|---|
| Policy held 1-5 years | | | |
| 30 to 39 | 2.16 | 0.13 | 16.6 to 1 |
| 40 to 49 | 1.49 | 0.58 | 2.6 to 1 |
| Policy held 6-10 years | | | |
| 30 to 39 | 1.93 | 0.26 | 7.4 to 1 |
| 40 to 49 | 1.79 | 0.23 | 7.8 to 1 |
| Policy held 11-15 years | | | |
| 30 to 39 | 2.12 | 0.24 | 8.8 to 1 |
| 40 to 49 | 1.06 | 0.35 | 3.0 to 1 |
| Policy held 1-15 years | | | |
| 30 to 39 | 1.97 | 0.22 | 9.0 to 1 |
| 40 to 49 | 1.49 | 0.34 | 4.4 to 1 |

Source: State Mutual Life Assurance Company of America, 1981.[2]

The implications of these data are staggering! In effect, premiums paid by employees less than age 50 for term-life coverage are used to underwrite death benefits for those who smoke.

The Medical Department of Dow Chemical Company tabulated deaths during a three-year period in a group of 1,352 employees who had worked for Dow at least fifteen years and whose smoking history was known. The major findings were: (1) the mortality rate for smokers was 3.5 times that of nonsmokers; and (2) the mortality rate for heavy smokers was 4.8 times that of nonsmokers.[9]

If smokers don't die before reaching retirement age, early disability may shorten their tenure of service—but not their compensation. In 1978, Chief Charles Rule of the Alexandria, Virginia, Fire Department ordered a complete ban on hiring smokers after he determined that 73 percent of those retiring due to disability in the previous five years were smokers who had developed heart or lung disease. He also calculated that each early retirement cost the taxpayers $300,000 more than a normal one.[10]

Putting a price tag on early disability and mortality among smokers is difficult, but Luce and Schweitzer estimated additional costs of $865 per smoker per year, adjusted to May 1980 dollars.[6]

## Fire Insurance

Organizations that forbid smoking on their premises should be in a good position to bargain for reduced accident and fire insurance premiums. Although this may be tough, Robin Woodward, a Seattle restaurateur who forbids smoking in her restaurants, negotiated a 25 percent reduction in her fire insurance premiums from Safeco Insurance Company. Farmers Insurance Group offers discounts on premiums to individual vehicle owners who don't smoke. An organization with a fleet of vehicles used only by nonsmokers should certainly seek a comparable discount.

## References

1. Kristein, M., "How Much Can Business Expect to Earn from Smoking Cessation?" presented to National Interagency Council on Smoking and Health's National Conference, Chicago, January 9, 1980, p. 3.

2. Naus et al., "Work Injuries and Smoking," in *Industrial Medicine and Surgery*, Oct. 1966, pp. 211-215. Yuste and DeGuevara, "Influencia del fumar en los accidentes laborales," *Med. y Seguridad del Trabajo*, Oct./Dec., 1973. Both cited in Kristein, op. cit., p. 4.

3. Cowell, M. J., and Hirst, B. L., "Mortality Differences Between Smokers and Nonsmokers," State Mutual Life Assurance Company, Worcester, Mass., 1981.

4. National Fire Protection Association, Boston, "Fires and Fire Losses Classified, 1974," *Fire Journal*, September 1974.

5. U. S. Fire Administration, Unpublished report to Department of Health and Human Services, derived from National Fire Incident Reporting System, Jan. 1985.

6. Luce B. R., and Schweitzer, S. O., "Smoking and Alcohol Abuse: A Comparison of Their Economic Consequences," *New England Journal of Medicine*, March 9, 1978, pp. 569-570.

7. Carlson, R., *Smoking or Health in New Jersey: A Progress Report on Making Nonsmoking the Norm*, New Jersey State Department of Health, Occupational Health Program, 1981, p. 26.

8. *Smoking and Health*, Report of the Surgeon General, U. S. Dept. of Health, Education and Welfare, 1979, table 6, pp. 2-18.

9. Fishbeck, W. A., Dow Chemical Company I.Q. Program. Dow Medical Department, Michigan Division, 1974.

10. Kelliher, E. V., "Fewer Workers Now Are Singing 'Smoke Gets in Your Eyes,'" *Wall Street Journal*, November 7, 1978, p. 1.

# 6

# Lower Maintenance Costs

At a seminar on smoking policies for the workplace, Robert Barnitt, Unigard Insurance vice president of personnel, read proudly from a letter the company had received from its cleaning service:

[Before your no-smoking policy was introduced] each smoker's desk had an ashtray to dump and clean, as well as ashes spread over the desktops and on the carpet around the desk. [Now] we don't have to dump and clean ashtrays. The dusting of desktops is easier. Carpets don't need to be edged or shampooed as often. Upholstered furniture is easier to keep clean. Windows don't get dirty as fast.

Because the work [has been] easier . . . [we have been] able to reduce the monthly service charge . . . $500 [of the reduction is] due to the no-smoking policy alone.[1]

Unigard shouldn't have been surprised. Routine maintenance always costs less in a smoke-free environment. Warren McPherson, former president of Seattle's Radar Electric, has reported:

Before we took over the upper floors—effectively doubling our square footage—our lone maintenance man was scrambling to keep the facility clean. We were planning to add *at least* one additional maintenance person. But then the smoking ban was implemented, and now my once-ragged maintenance man easily maintains twice the floor space, and pokes around like a Maytag repairman looking for other chores to keep himself busy.

In today's dollars, the decision is saving me $30,000 per year on cleaning costs alone. And if you want to spread that over the number of smokers that once worked here, it comes to about $750 per smoker per year.

McPherson explained the reduction in cleaning costs this way:

> It's easy to see why. The floors needed sweeping three times
> a week during the smoking days, compared to once a week
> now. Twice every month we needed to completely wash all
> exposed surfaces in the showroom and office area—win-
> dows, display cases, and so forth. Now the wash is *once a*
> *year*. Employees and customers really appreciate a spotless
> environment and that in itself has helped with the main-
> tenance problem. It's funny, but people seem to be looking
> for a mat to wipe their feet on as they enter. They did not
> do that when the floor was littered with butts and ashes.
> And then, of course, there were the ashtrays, which always
> needed emptying.

The experiences of Unigard Insurance and Radar Electric are
typical. Cheerful stories have also been reported by Boyd Coffee,
Riviera Motors, Continental Telephone of the Northwest, the
Austad Company, and many others. Merle Norman Cosmetics,
for example, estimated that it saved $13,500 on janitorial services
for 1976 in a small plant where smoking was banned.[2]

## Damage to Equipment and Furniture

Routine housekeeping expenses can be a small annoyance compared to the cost of smoke damage to delicate equipment and
fine furnishings. Some of the first electronic companies had
expensive precision equipment ruined because they allowed smoking nearby. Now smoking is no more tolerated around such in-
struments than it would be around oxygen tanks.

How unfortunate that many companies still regard their
equipment more highly than their employees! New Jersey Bell,
for example, lost a lawsuit because the company banned smoking
around its sensitive equipment but not around the nonsmoker

who suffered eye abrasions and nosebleeds due to tobacco smoke.[3] Damage to delicate equipment is often the tip of an iceberg. Less delicate equipment, such as typewriters, are often damaged at a slower rate by the tars and ashes from smoking. The connection is often overlooked because the damage is not as immediate or disruptive.

Furniture is often burned by cigarettes, and office furnishings, such as carpeting and draperies, are often replaced for the same reason. Notes McPherson:

> Eliminate smoking and you eliminate the primary agent of depreciation—thereby extending the useful life of the average piece of furniture by a sizable factor. I expect to get at least three times the life from our office furniture today, and we select classic styles with that in mind.

Concurring with McPherson's observation is Gardner Kent, owner of the imaginative Green Tortoise bus system based in San Francisco. Each diesel coach has been remodeled by replacing the uncomfortable upright seats. The rear half of each coach is upholstered with foam for comfortable stretching and sleeping. In the front half are tables and couches. After the first few trips in 1973, when the system began, Kent banned smoking "because of the emotional trauma" one of his nonsmoking passengers went through. The foam covers no longer needed to be replaced—at a cost of $20 apiece—because of burns.

## Redecorating Costs

Another way organizations have coped with the residue of smoking is by covering it with paint. McPherson says:

> There is simply no need for repainting, except on a very infrequent basis—say every ten years just to change the

color. I'm convinced that repainting is necessitated by the
accumulation of tobacco smoke, not by the paint fading.
No tobacco smoke, no dingy walls.

Lyndon Sanders, who owns the Nonsmokers' Inn in Dallas
and the Dollar Inn in Albuquerque, says his 110 smoking rooms
in the Dollar Inn need painting five times more often than the
nonsmoking ones. He also reports that rooms used by smokers
take 40 percent longer to clean thoroughly. Sanders reaps $1,000
more profit per year from each nonsmoking room because his
cleaning costs are lower.

In 1984, National Car Rental Systems in Minneapolis budget-
ed about $50,000 to clean its ceilings, which reflect the indirect
lighting in its four-story sealed building. Keeping the ceilings
clean is essential for good lighting. Mike Jones, a financial analyst
who served on the company's committee to develop a no-smoking
policy, said that the "vast majority of the light reduction is caused
by yellow stains from smoking . . . It's obvious that the ceilings
with the highest incidence of smoking below get more cleaning
than the rest." Jones said that 455 (35%) of the 1,300 employees
smoke, which means that the cost of cleaning the ceilings is $120
per year per smoker. Even if a smoking ban only halved the
annual cleaning bill—and it would probably reduce it more than
that—the other $25,000 could be spent for in-house health pro-
motion, a far wiser investment for the company.

## Ventilation Costs

Why not just ventilate the smoke and impurities to the outside?
The cost would be enormous. According to an article in *Energy
Management Reports*:

> Ventilation rates must be increased 7 to 10 times above
> what is needed where there is no smoking, and many

ventilation systems are unable to deliver a relatively pure environment [when there is] cigarette smoke at any level . . . and even if they could, the cost would be astronomical.[4]

Standard 61-81 of the American Society of Heating, Refrigerating and Air Conditioning Engineers (ASHRAE), which covers "Ventilation for Acceptable Air Quality," requires a turnover of 5 cubic feet per minute per person in no-smoking areas. Where smoking is allowed, the minimum turnover required is 20 cubic feet per minute. Based on these figures, three engineers calculated that the extra heat needed in a climate with 5,000 heating degree days per year was 1.95 million BTUs per smoker. Assuming that oil costs $1.20/gallon and is burned at 60 percent efficiency, the additional cost per smoker per year would be $27.57.[5]

Keep in mind that these costs reflect air-quality standards that allow concessions to the smoking minority. According to Richard Duffee, director of odor technology at TRC Environmental Consultants, "The levels of particulate matter in office buildings where smoking is allowed are 10 to 100 times higher than the allowable limits set for outside air. And where smoking is allowed, no amount of ventilation gets rid of that odor."[6]

Jet planes also have forced ventilation systems. Bob Fox, formerly logistics coordinator for the customer support group at Boeing in Seattle, believes that smoking bans will save airlines money. In 1981, in a letter to chairman Marvin Cohen of the Civil Aeronautics Board, Fox explained:

Smoke-laden air flows in contact with the cold fuselage skin, where the tobacco tars solidify and stick to the skin and structure, [causing a] sticky mess, which accumulates at the rate of 200 pounds per year on a 747.

At the present rate of 5 pounds of fuel burn per hour for each 100 pounds of weight, a 10-year-old 747 uses 52,000 gallons of fuel per year to haul the ton of tobacco tar around.

Clearly, smoke-free policies can immediately reduce the cost of maintenance, property damage, and ventilation. And, like the costs of insurance against fires and industrial accidents, smoking's costs do not generate more customers or profits.

# References

1. Corporate Smoking Policies Seminar, Bellevue, Washington, July 21, 1982, sponsored by the American Lung Association of Washington.

2. "Fewer Workers Now Are Singing 'Smoke Gets in Your Eyes,'" *Wall Street Journal*, November 7, 1978, pp. 1, 33.

3. Shimp, Donna M., Blumrosen, Alfred W., and Finifter, Stuart B., *How To Protect Your Health At Work*, published by Environmental Improvement Associates, 109-111 Chestnut Street, Salem, NJ 08079.

4. *Energy Management Reports*, July 1980, p. 8.

5. "Achtung! Smoking May Be Hazardous to Your Energy Budget," *Solar Age*, June 1982, p. 12.

6. Kaufman, J.: "Indoor Air Pollution Raises Risks for People in New Office Buildings," *Wall Street Journal*, July 16, 1980, p. 35.

# 7

# The Cost of Passive Smoking

Perhaps you've heard about canaries in coal mines. Those caged birds extended the miners' senses by providing an early warning signal that dangerous gases were flowing through the mine. When the birds dropped dead, the miners scrambled out of the tunnels to avoid a similar fate.

Nonsmokers, however, don't need canaries to learn about the quality of their air at work. Their senses serve that function. Irritated noses and lungs are normal reactions to irritants and poisons in the air. But symptoms like this, which warn that corrective action is needed, often fail to get the attention they deserve. Smokers and management often dismiss these reactions by saying they are personality quirks or that complainers are "hypersensitive." It is true that some nonsmokers are so sensitive that smoke can trigger asthma attacks or other serious problems. But defensive use of the word "hypersensitive" is merely a smokescreen for inaction.

## Survey Results

Group surveys have confirmed what individual nonsmokers have known all along—that smoke in the workplace causes discomfort. For example, a survey of more than 20,000 workers at the Social Security Administration Headquarters in Washington, D.C., found that nonsmokers exposed to tobacco smoke complained about the following: conjunctival irritation (47.7%); nasal discomfort (37.7%); coughing, sore throat, or sneezing (30.3%); and difficulty digesting food (19.4%).[1]

Of 472 workers surveyed in Switzerland, 43 percent (of

smokers and nonsmokers) reported eye irritations from tobacco smoke. The percentage for smokers was 27 percent and that for nonsmokers was 58 percent.[2].

In a survey conducted in Toronto, 75 percent of the non-smokers said tobacco smoke annoyed them in restaurants and 53 percent said that it bothered them in their offices.[3] Pacific Telephone and Telegraph in California surveyed 2,942 of its more than 100,000 employees. Table 7:1 lists the ways smoking and nonsmoking employees of this company said they were bothered by tobacco smoke.

Table 7:1. Worker Complaints About Tobacco Smoke

| Problem Reported | % Employees |
| --- | --- |
| Eye irritation | 66 |
| Clothes and hair smell | 61 |
| Concern about long-term effects on health | 58 |
| Coughing and/or throat irritation | 36 |
| Headaches | 32 |
| Interference with work performance | 27 |
| Unpleasant odor | 6 |
| Sinus or nasal problems | 5 |
| Breathing problems | 4 |
| Nausea | 3 |
| Dirt from ashes and residue | 3 |
| Inadequate ventilation | 3 |
| Other reasons | 6 |

Source: *Employee Smoking Study,* Project #82-63, Corporate Research Division, Human Resources Administration, Pacific Telephone and Telegraph, January 1983.

Constant annoyances can create stress and impair a person's ability to work. In the Social Security Administration survey, nonsmokers reported three main responses to smoke: 51.2 percent said they had difficulty working near a smoker; 20.6 percent said they had difficulty concentrating on work; and 13.5 percent said they had difficulty producing work.

A nationwide survey sponsored by Honeywell Technalysis and conducted by Public Attitudes of New York found that cigarette smoke bothered 54 percent of the 600 workers surveyed and that 53 percent said that cleaner air would make their organizations more productive. The symptoms mentioned most frequently were: tired or sleepy feelings (56%), nasal congestion (45%), eye irritations (41%), breathing difficulties (40%), and headaches (35%).[4]

According to James E. Woods, Ph.D., senior staff scientist for Honeywell's Corporate Physical Sciences Center: "The most common indoor-air problem is cigarette smoke. It contains more than 3,000 contaminants that can cause respiratory irritations and serious health problems for nonsmokers as well as smokers."

After a comprehensive review of research, Roy J. Shepard concluded in *The Risks of Passive Smoking* [Oxford University Press, New York, 1982] that "eye irritation seems to be the main complaint during passive exposure to cigarette smoke" and that "more annoyance is caused by smouldering cigarettes than by active smoking."

Before smoking was banned in the newsroom of the *Bellevue Journal American* in Bellevue, Washington, a flurry of memoranda was exchanged between smokers, nonsmokers, and executive editor Frank Wetzel. Nonsmokers finally sent Wetzel a petition asking him to take a stand on the issue. Many reported daily headaches, watery eyes, and sore throats. The marked improvement after the ban is illustrated by these memoranda:

> *Nonsmoker #1:* "I want you to know how good it is to be able to sit down to work and not have to spend half the time concentrating on how to get away from the smoke . . . I was finding it just about impossible to work, and now that's no longer the case. I also appreciate not having to go home every afternoon with a headache."

*Nonsmoker #2:* "It is wonderful to be able to take a deep breath without feeling like I have to cough. And though I'm tired after a day's work, I'm not nearly as tired and deflated as I was when the newsroom was thick and heavy with smoke. And no headaches or smarting eyes. WOW. Another benefit—I've noticed the newsroom smell is going away."

## Documentation of Illness

In 1985, Californians for Nonsmokers' Rights summarized the hazards of tobacco smoke in "Tobacco Smoke and the Nonsmoker," a brochure prepared from more than 500 specific publications. Among other things, the brochure notes:

● Secondhand smoke, like all tobacco smoke, contains thousands of toxic chemicals, including cyanide, arsenic, formaldehyde, carbon monoxide, and ammonia. Tobacco-related cancer-causing chemicals appear in the urine of nonsmokers exposed to cigarette smoke.

● Breathing secondhand smoke significantly increases the risk of developing cancer. Several studies report that nonsmoking wives of smoking husbands face 2 to 3 times the risk of lung cancer as nonsmoking wives of nonsmokers. Other studies have revealed that involuntary smoking is associated with doubling of the overall cancer risk.

● Breathing secondhand smoke for extended periods causes other diseases in healthy nonsmokers.

● Children of smoking parents have more respiratory illnesses and allergic manifestations than similar children of nonsmoking parents. The illnesses appear to be dose-related; that is, when both parents smoke, the children have more respiratory illnesses than if only one parent smokes.

● Teenagers have impaired lung function when their parents

smoke; this effect is independent of and additive to the effects of any smoking by the teenagers themselves.

● Lung function of adult nonsmokers who work in smoky offices for years becomes impaired to the same extent as that of light smokers. Nonsmokers working in smoke-free offices have better lung function than the light smokers and passive smokers.

## Cost to the Company

Researchers J. L. Repace and A. H. Lowrey, who reviewed 14 epidemiological studies, found that all but one showed evidence that nonsmokers exposed to cigarette smoke had an elevated risk of cancer. They estimate that ambient smoke causes between 500 and 5,000 lung cancer deaths per year in the United States, and some of the studies suggest that the risk is doubled. They note that mainstream tobacco smoke (the smoke inhaled by the smoker) is a potent carcinogen. They also estimate that mortality from passive smoking is many times that of cancer-causing chemicals regulated by the Federal Clean Air Act.[5] Researcher Peter Fong says "ambient smoke is estimated to cause an excess of deaths between 50,000 and 10,000 a year in a population of 220,000,000."[6]

Based on data from the R. J. Reynolds Tobacco Company, R. Fritz Hafer, Ph.D., and Floyd Frost, Ph.D., calculated that the average passive smoker inhales between 1/1000 to 1/100 cigarette per hour. Assuming 1/500 cigarette per hour for 12 hours by 73 million passive smokers in the United States, Hafer and Frost figure this "amounts to 1,750,000 cigarettes *per day* 'smoked' by nonsmokers . . . Concerns over health effects from passive cigarette smoke are hardly a myth . . . passive cigarette smoke is likely the most dangerous air pollutant we face today."[7]

White and Froeb have concluded that nonsmokers who work in a smoky environment have about the same risk of impairment in the small airways of the lungs as do smokers who do not inhale and smokers who inhale between one and ten cigarettes per day.[8] An earlier study by Snyder and Stellman concluded that the latter group suffers approximately one-fifth the damage of normal smokers.[9] As noted in Table 7:2, each smoker could be increasing an employer's cost for every nonsmoker on the payroll by about $243. Because each smoker puts approximately two nonsmokers at risk, the additional payroll cost to an organization because of nonsmokers' poor health from tobacco smoke would be about $486 per smoker per year. And, as noted in Chapter 3, when moking-induced absenteeism reduces the number of billable hours per week, the loss can be substantially greater.

Table 7:2. Extra Annual Costs Per Nonsmoker in Smoky Workplace

| Expense Category | Cost Per Smoker | Cost Per Nonsmoker |
|---|---|---|
| Absenteeism | $ 220 | $ 44 |
| Health insurance | 230 | 46 |
| Illness and early death | 765 | 153 |
| Total | $1,215 | $243 |

In 1981 dollars, based on total personnel cost of $20,000 per employee. Source: Weis, W. L., "Can You Afford to Hire Smokers?" *Personnel Administrator*, May 1981.

## Psychological Harm

Nonsmokers suffer not only from physical afflictions attributable to smoke, but from psychological ones as well. In the Social Security Administration survey mentioned above, employees were asked about their mental reactions to smoke. Among nonsmokers, 24.2 percent reported frustration because of the smoky nuisance and 22.3 percent reported hostility toward smokers and management. Consider the morale of this nonsmoker after getting this memorandum from a smoker he had asked not to smoke:

The very law of dispersion indicates that generally what you are smelling is not tobacco smoke but something else.

There are few times that I enter your den of purity with a foul smelling fag in my grubby paw and in the majority of those times the smoke emitting from my long white stick is dispersed and you are picking up only remnants with your long enduring nose.

It is my hope that only the methane gas reaches and accumulates in your big toe and I am confident that you shall surely die of cancer from this city long before you do from my Marlboro.

As reported in the next chapter, these frustrations and hostilities erode employee morale, the desire to do good work, and organizational loyalty. Exactly what percentage of nonsmokers quit their jobs because of smoky working conditions is difficult to assess. But turnover is definitely increased. The nonsmoker who received the memo above quit after nine months on the payroll—after $2,000 in addition to his professional salary was spent for his training.

# References

1. "Smoking on the Job: The Controversy Heats Up," *Occupational Health and Safety*, Jan./Feb. 1979.

2. "Passive Smoking at Work," *International Archives of Environmental Health* 47:209-221, 1980.

3. Shephard, R. J., and Labarre, R., "Attitudes Toward Smoking and Cigarette Smoke," *The Toronto Survey*, University of Toronto, 1976.

4. *Indoor Air Quality: A National Survey of Office Worker Attitudes*, sponsored by Honeywell Technalysis, Honeywell, Inc., Minneapolis, MN 55408, February 1985.

5. Repace, J. L., and Lowrey, A. H., "A Quantitative Estimate of Nonsmokers' Lung Cancer Risk from Passive Smoking," *Environmental International* 11(1):3-22, 1985.

6. "The Hazard of Cigarette Smoke to Nonsmokers," *Journal of Bio-*

*logical Physics* 10:65, 1982.

7. "Passive Smoking: Myth or Reality," *Morbidity Report,* Office of Public Health Laboratories and Epidemiology, State of Washington, Olympia, Washington, May-June 1984, vol. 1, no. 2, p. 4.

8. "Small-Airways Dysfunction in Nonsmokers Chronically Exposed to Tobacco Smoke," *New England Journal of Medicine,* March 27, 1980, pp. 720-723.

9. "Comparative Epidemiology of Tobacco-Related Cancers," *Cancer Research,* December 1977, pp. 4608-4622.

# 8

# Nonsmoker Frustration

Bad air has become the number-one complaint about indoor work. Across the country, workers are filing complaints about its main cause: smoking on the job. Burning eyes and lungs, asthma attacks, and swollen sinuses are being reported.

Medical research is turning up additional hazards. One study has concluded that prolonged exposure to secondhand smoke produces small airway dysfunction in the lungs. Other studies point to a greater risk of nonsmokers' contracting lung cancer by exposure to tobacco smoke. Even "bingo brain"—a condition of mental confusion described in the June 1, 1982, *Canadian Medical Association Journal*—has been attributed to high concentrations of carbon monoxide in a smoky bingo parlor.

## Many Smokers are Indifferent

Fired up by burning senses, armed with medical evidence, nonsmokers are speaking up at work, hoping to clear the air. Many smokers don't seem to care. When Gary Fox worked at the Boeing Aerospace Center in Kent, Washington, he complained to his boss for a week that smoke bothered him. The boss (a smoker) finally said, "The desk you're working on has always been there. It's always going to be there, and that's where you're going to work." Frustrated, Fox called the Environmental Protection Agency. "I called . . . on a Friday. Monday, they sent two people out. Wednesday, I was transferred to another building," Fox reported. "They just wanted me out of their hair."

Roger Maldonado, a bus driver for Jefferson County Transit, posted a sign in the Port Townsend, Washington, drivers' lounge asking smokers to please clean their ashtrays. The request was usually ignored; to get relief from the resultant odor of smoke, Maldonado emptied the ashtrays himself.

## Persistence May Be Required

For some nonsmokers the battle can be long, hard, and frustrating. A few years ago, Janet Schmirler worked for Pacific Bank in Seattle (now called First Interstate). "In the mortgage division I was so physically uncomfortable that it was really difficult to do my work," she has reported. Her smoking co-workers were indifferent and began avoiding her. When appeals to her supervisor did nothing, she wrote to the bank's personnel office.

The official response acknowledged that "some smokers need to be reminded to show consideration for those who may be adversely affected by cigarette smoke." But it did nothing more than refer her back to the supervisor, who, according to Schmirler, was a heavy smoker and unwilling to restrict smoking for fear that other smokers would spend too much time away from the job on "smoke breaks." The lack of a policy to protect nonsmokers at the bank, she heard, was due to top management's fear of losing middle managers, many of whom smoked.

After three months in the mortgage division, Schmirler secured a transfer to a better ventilated department. But the presence of smoke still bothered her enough that she began looking for a smoke-free job. After a total of 13 months at the bank, she found one and took it.

Schmirler's story is short compared to that of Eldon Ball, whose struggle with King County, Washington, began in 1979. Two months after Ball began work in the county administration building, he posted a sign near his desk that said: "Our policy is no smoking. Thanks for your cooperation."

"My boss didn't like its saying 'our policy,' kind of indicating it was county policy," says Ball. "So I ran a line through 'our' and wrote 'my.'" Later he put up small signs to encourage cooperation. "These things would turn up missing," he said. "So I started taping them down. People would cut the tape and remove them."

In February 1981, Ball wrote a memo to his boss asking that the office be made a nonsmoking area and citing the health reasons for this request. Soon afterward, "when it got around that I had done this, I felt a certain pressure from the smokers in the group." Other nonsmokers, Ball claims, were afraid to join with him in his request for fear of retaliation from the boss, fellow workers, or supervisors.

During the next six months Ball submitted three more memos. Anonymously, word reached the boss that Ball was spending county time making phone calls to other governmental agencies for help. Ball says he made these calls on his lunch hour. The telephone was removed from his desk, even though its use was an integral part of his job.

In 1983, when the ventilation system in his office wasn't working to capacity, Ball suggested that smoking be prohibited until full ventilation was restored. The boss's reply, according to Ball, went something like, "Well, when you become county executive, you can implement that."

In June 1983, Ball was moved into another room, just ten feet around a wall from a smoker. Taking direct action, he bought a 6-inch fan to divert the smoke. In the meantime, he continued to complain through memos. Toward the end of 1983, he heard from the Seattle-King County Health Department that a policy on smoking was under consideration. In May 1984, he heard on the radio that the county executive was about to order managers to develop smoking restrictions for their own divisions—a weak policy, according to Ball, because his manager was a chain-smoker.

Ball then sent a memorandum to his boss suggesting a three-stage process: (1) starting immediately, smoking would be confined to smokers' desks; (2) in November, the day of the Great American Smokeout, smoking would be confined to the coffee room during breaks; and (3) on May 15, 1985, smoking would be banned completely.

Ball got no reply. Instead, it was announced that a smoking policy development committee had been appointed, with him included. Surprisingly, Ball found himself attacked for his activism by nonsmoking committee members. Meanwhile, some smokers in the office devised a floor plan to segregate smokers and nonsmokers. "The Committee's attitude," says Ball, "was not how we could protect the nonsmokers, but how we could preserve the smoking."

The Committee's report went to the boss about June 1, 1984. On July 9, a policy dated July 2 was posted but not distributed to the employees. The policy had three provisions that Ball had suggested 39 months before:

1. Walking with a lit cigarette/cigar or pipe through nonsmoking areas will be prohibited.

2. Burning material shall not be left unattended in designated smoking areas.

3. When available, "smoke eaters" will be utilized in designated smoking areas. Smokers using these items should change the filters regularly and turn them off at the end of each shift.

On July 20, Ball was moved to a different section to help with extra work. He was placed a few feet from an "overly hostile" smoker who was verbally abusive. A five-foot-high portable screen was put up to help confine the smoke, but it was not effective. Ball turned again to his small fan, propping it on top of a file cabinet, which also served as a bulletin board for articles on secondhand smoke. Mysteriously, the articles disappeared and the fan would be found pointing in other directions.

In October, Ball asked the boss when no-smoking signs, a part of the county executive's mandate, would be posted. "Thank you for not smoking" signs went up, but in Ball's room, above a smoker, was posted the sign "Smoking permitted in this room." When 1984 Christmas spirits rolled around, someone remembered Ball. Taped to his "Letter Licker" was an obscene message.

On January 25, 1985, Ball went to a doctor for a shoulder injury from skiing. Seizing the opportunity, Ball explained the tobacco smoke problem at work. The doctor gave him a simple letter stating that "Mr. Ball has had recurring eye and nose problems due to cigarette smoke. He should have a smoke-free environment." After Ball submitted the doctor's statement, the personnel department told him the matter was under his boss's jurisdiction. At the end of February, Ball met with the boss's administrative assistant, who offered to move Ball into a semi-enclosed area off a hallway in which smoking was supposedly prohibited. However, this particular hall was used in part to lay out maps, and frequently those looking at the maps would perch their lit cigarettes near Ball's doorway.

By this time other nonsmokers in Ball's office had begun to extinguish cigarettes left burning in ashtrays with drops of water from an eye dropper. In April 1985, while passing a smoker in the hall who was looking at a map with Ball's boss, Ball dipped his fingers into a cup of water and let water drip onto the cigarette. His boss was outraged and issued a letter of reprimand:

> During the [map] review, Mr. [X], myself, and the maps were sprinkled with water. The water was thrown by you in an apparent reaction to a cigarette. . . . I directed you to never again take an action of that sort. . . . This is a letter of reprimand regarding the incident. It is my understanding that any future incidents of a similar nature by you could be grounds for disciplinary action up to and including termination.

Another recent example of a smoker's extreme lack of consideration occurred on CBS-TV's *Nightwatch* program on February 18, 1985, when Action on Smoking and Health (ASH) director John F. Banzhaff III debated a Fordham University law professor about nonsmokers' rights. As the program began, the professor smoked a cigar despite objections from Banzhaff and several members of the audience who said they were getting sick. After pointing out that the professor was violating the Washington, D.C., fire regulations and commenting on his extreme lack of courtesy, Banzhaff finally dumped a small cup of water on the cigar, extinguishing it. What followed was described by him in the March issue of ASH's *Smoking and Health Review*:

> While screaming at me for having been "violent," my opponent then threw an ashtray at my head and hit me at least three times in the chest. I, of course, had only used the least amount of force necessary under the circumstances—force directed against the cigar rather than him.
>
> NEITHER ASH NOR I RECOMMEND using such tactics with smokers, except in extreme circumstances such as this. Indeed, in more than 12 years of debates, and a lifetime of living around smokers, I have never found it necessary before. But in some limited situations, the law does give nonsmokers the right to defend themselves from smoke.

## Other Strategies

Unlike Ball and Banzhaff, Seattle City Light worker Louise McClellan tried fighting smoke with smoke—by burning incense. For her attempts to sweeten the foul air she received a reprimand from the management.

That smokers can burn tobacco, but nonsmokers can't burn incense, is just one example of how smokers are given "protection," according to Ramona Hensrude, past president of Fresh Air for Nonsmokers, Washington state's nonsmokers' rights coalition. "They are given every possible opportunity to continue their habit without being responsible for the breathing air they pollute for others." When smoky air is imposed on nonsmokers, they can become "demoralized because of the frustration and the delay in making the workplace a healthier environment. Then nonsmokers lose respect for supervisors and co-workers." Hensrude notes further that management's indifference and inertia often occur because they don't understand the issue, taking it personally instead of seeing it as a health issue.

The mental suffering of nonsmokers resulting from a sense of helplessness is not easily remedied. Cornelius J. Peck, professor of law at the University of Washington, says there may be little recourse for nonsmokers who are harassed and taunted by management or co-workers. The employment-at-will doctrine, he explains, allows employers to hire and fire as they please unless union contracts or civil service regulations intervene. Without such protection, vocal nonsmokers run the risk of losing their jobs because the "employer can be just exactly that arbitrary."

Peck sees possible recourse through laws protecting the handicapped. He suggests that, if nonsmokers could be designated handicapped persons because of their allergies, they could bring action under state statutes that govern discrimination in employment. But that's a difficult process and wouldn't work for "mere annoyance," Peck adds.

Depending on the kind of smoke a male boss or co-worker blows at a female employee, Peck entertains another possibility. "If he is blowing cigar or pipe smoke and saying you have to take it, then I see glimmering chances of turning this into some kind of sexual harassment case. It's men who smoke cigars. It is therefore a male thing to do, to harass women."

Another option Peck sees is the tort of outrage—a civil action. But he concedes that the "smoking harassment has to be severe enough to be categorized as outrage or intentionally causing severe emotional distress. Extreme conduct is necessary to get any kind of tort action."

One might think the legal system itself would be gracious and fair. Not so, says E. J. Metcalf, who was called for jury duty in the King County Superior Court. During jury selection, Metcalf, who has only 35 percent lung capacity because of emphysema, was seated near smokers. Finally she asked a smoker to move. Metcalf recounted the subsequent unpleasantries in a letter to the court:

> A female court clerk, herself smoking, ordered me to the other end of the room and very rudely said, "Lady, you move to the other end." By this time I was devastated and felt like I was a criminal on trial. This was a very bad experience for me and I never want to be called for jury duty again.

## The Price of Indifference

Until management takes a stand for health and profit, nonsmokers will find ways to express their discomfort and disdain for polluted air. Adverse reactions to smoke are often perceived by management as mere annoyances that can be ignored. "Mere annoyances" or not, they are real concerns to nonsmokers.

Management and smokers seem to overlook their own culpability in this denial process and the interpersonal problems that ensue. Rather than deal constructively by promoting health, which might entail a ban on smoking, they try to justify their unhealthy behaviors.

When normal reactions to unhealthy conditions are consistently denied and persecuted, employee morale and loyalty fade. Employees unable to leave such an appalling situation may become bitter, inefficient workers, perhaps determined to exact revenge on the employer. Others—very likely the brighter and better workers who want to feel good about what they do—will seek work in a healthy environment. Who loses in the end? The uncaring employer.

# 9

# Smoking Bans Are Legal!

The belief that employees don't mind working in smoke may well have been conjured up in a smoke-filled corporate boardroom. But it is clear that most workers disagree. In a survey of 3,000 Hawaii State Health Department employees, 77 percent (including 65.6 percent of the smokers) favored no-smoking sections in common work areas and 44 percent (17.8 percent of the smokers) favored a total ban.[1]

When 21,000 employees at the headquarters of the Social Security Administration were surveyed, 70 percent returned their questionnaires. Asked to choose among three smoke-free strategies, 47.3 percent (37.9 percent of the smokers) favored designated smoking areas away from the work area; 27.1 percent (49.5 percent of the smokers) asked for separate work areas; and 25.6 percent (12.6 percent of the smokers) wanted a ban.[2]

In a survey of 3,000 workers of East German Railways (Deutsche Reichsbahn), 79 percent of the smokers and 95 percent of the nonsmokers said smoking in work areas should not be considered a right if the majority of the workers were nonsmokers. In addition, 70 percent of the smokers and 93 percent of the nonsmokers agreed that a smoker should refrain from smoking when asked to do so by a nonsmoker.[3]

In Zurich, Switzerland, 472 people working in 41 rooms were asked their preference between a clean or smoky work environment. Just under half, 49 percent, favored complete prohibition of smoking, and 70 percent wanted at least separate work areas for smokers and nonsmokers.[4]

In yet another survey, reported in the *Wall Street Journal*, 51 percent of 1,004 office workers thought smoking should be restricted to certain areas.[5]

Viewers of KIRO-TV's *Newsline* were asked the single question, "Should separate smoking and nonsmoking areas be required in private offices?" Of the 431 who called in, 88 percent said yes.[6]

Pacific Telephone and Telegraph in San Francisco surveyed nearly 3,000 of its employees and found that 51 percent of the smokers and 80 percent of the nonsmokers favored some sort of smoking restriction.[7]

In Seattle, Pacific Northwest Bell Telephone's survey of 1,334 employees also confirmed workers' desires for smoke-free policies. In this survey, 77 percent (57 percent of the smokers) said they wanted a policy for the immediate work area. Fifty-two percent of the nonsmokers and 49 percent of the smokers preferred designated smoking and nonsmoking areas as the policy. Thirty-two percent of the nonsmokers and 9 percent of the smokers preferred a total smoking ban.[8]

Seventy-seven percent of 311 nonsmoking readers who responded to a recent survey by the British magazine *Ms. London* said that smoke in the workplace was "extremely unpleasant."

A survey of 371 Seattle-area managers showed that 88 percent favored smoking restrictions at work. Fifty-four percent favored separate smoking and nonsmoking areas, and 34 percent favored a total ban.[9]

The fact that most workers prefer to work in air unpolluted by tobacco smoke should reassure employers that restrictions—or a ban—will be welcomed by the majority of employees, including many smokers who want to stop smoking. Employers who assume otherwise are setting themselves up for complaints, low morale, and even legal action by nonsmokers.

Surveys after smoking has been restricted confirm employee preference for smoke-free workplaces. In April 1984, Group Health Cooperative of Puget Sound banned smoking on the premises except for a few designated rooms for patients. The policy affected some 6,000 employees. Six months later a random sample of 687 employees was asked about the new policy. Eighty-five percent approved, 11 percent more than before the policy went into effect.

## Employer Reluctance

But even with the majority of the workforce favoring smoking restrictions on the job, many employers continue to ignore complaints about smoke for fear of upsetting or losing employees. Some of these employers pretend the smoking issue is just a simple squabble between smokers and nonsmokers that will go away. Or that nonsmokers need something to complain about. Or that the tension will end when all the smokers quit smoking. Other employers simply don't know what to do.

This stance is odd considering the "employment-at-will" doctrine that employees serve at the employer's pleasure. No law gives employees the right to wear shorts to work in opposition to company policy, or to smoke on the job, with or without company policy. Basically, the employer still pulls the strings.

At the 1983 Conference on Smoking and Health sponsored by the American Lung Association of Washington, Timothy J. Lowenberg, a labor attorney and professor at the University of Puget Sound School of Law, told a group of employers:

> Many employers have justified their inaction by claiming that they are under an affirmative obligation to protect the rights of smokers as much as nonsmokers and that in considering a smoking policy for their company they must, as a matter of law, try to strike a balance between the rights of smokers and the rights of nonsmokers. This fence-straddling policy is a fallacious one because it rests upon a fundamentally unsound premise; that is, that smokers have a legal right to smoke in the workplace.

Several court cases directly address the false presumption that an employer must permit smoking at the workplace. For example, in *Parodi vs. Merit Systems Protection Board*, the U.S. 9th Circuit Court of Appeals in San Francisco ordered Ms. Parodi's employer (the federal government) to offer her either "suitable employment in a safe environment" (one totally smoke-free) or to pay her $500 a month plus allowable increases until she reaches retirement age. In addition, she was awarded approximately $20,000 in retroactive disability pay.[10] (Ms. Parodi, who was 47 when the decision was reached on October 21, 1982, suffered from asthmatic bronchitis.)

In *Smith vs. Western Electric Co.*, the Missouri Court of Appeals held that a nonsmoker made ill by secondhand smoke at work had the right to require his employer to provide a totally smoke-free work environment. The defendant in the case, a private employer, held the widely shared but untenable view that smokers have "rights" to smoke at the workplace.[11] In an amicus brief submitted on behalf of the Clean Indoor Air Foundation of Massachusetts and Environmental Improvement Associates of New Jersey, Suffolk University Law School professor Alvan Brody offered important advice to employers:

> The fundamental fallacy of the defendant's "smoking poli-cy" is its assumption that smoking employees have a "right" to smoke at their desks, even if it means smoking into the air other employees nearby must breathe. Where does this "right" come from? It is not conferred by the common law or by statute. On the contrary, the common law from its earliest origins established a contrary principle—that every-one has a right to the integrity of his body, a right not to have his body unnecessarily intruded upon by others. Under basic common law principles a smoker's "right" to smoke stops when his smoke intrudes upon another's body without his consent or acquiescence. As Bernard Shaw observed, "A smoker and a nonsmoker cannot be equally free in the

same railroad carriage."

Put another way, the defendant's "smoking policy" is, at bottom, that the interest of those employees who have become accustomed to smoking at their desks should be catered to by putting the plaintiff and the defendant's other nonsmoking employees at risk of their health.

If some of the defendant's employees started pinching other employees, or slapping them on the buttocks, or spitting on their sleeves, or spraying ammonia about in small quantities, and the defendant knew of the practice and knew that the victims objected to it, surely the defendant would not be talking about pinchers' rights, or slappers' rights, or spitters' rights or sprayers' rights. It would put a stop to such practices, and quickly.

Still, some employers are indifferent to the discomfort of nonsmokers. The following request was submitted by an employee at City Light, Seattle, Washington, through the company's formal suggestion procedure:

> I would like smoking banned in the office or allowed only during breaks in specially designated areas. What is a nonsmoker supposed to do other than suffer in silence?

The official response was:

> City Light is in full compliance with the rules set forth in WAC (Washington Administrative Code) 248-152-010, adopted March 12, 1975, by the state legislature. While the statute does not presently authorize measures as stringent as those suggested, City Light has undertaken a number of projects for the betterment of its employees. These include: 1) a separate cafeteria nonsmoking area; 2) no smoking in the library; 3) a trial project with three pool vehicles set aside for nonsmokers.
>
> Concern for good health is appreciated. City Light will

continue to make improvements to the office environment wherever resources permit or whenever mandated.

Of course this answer totally evaded the employee's concerns about smoke at the workplace. All three of the company's actions ignored the main issue of smoking where work is to be done—and the effects of the smoke on the workers. Attorney Lowenberg says:

> The absence of a meaningful general statute in any given state should be of no solace to the employer who wants to avoid taking a position . . . because there is a rapidly developing body of law which is resulting in the award of disability benefits, unemployment compensation benefits, injunctive relief and other judicial remedies on discrimination and handicapped theories to workers who suffer the ill effects of exposure to smoke in the workplace.

## Dealing with Unions

The presence of unions can cause employers to think smoking restrictions are impossible. But Cornelius J. Peck, a University of Washington law professor, has commented that unions are not a barrier if an employer proceeds according to union contracts and National Labor Relations Board (NLRB) rules:

> If the employer is going to announce a new no-smoking rule, I think this would be treated by the National Labor Relations Board as a change in working conditions and therefore something that would have to be bargained out with the union before the rule could be put into effect.[12]

When the Johns-Manville Company attempted to introduce a no-smoking rule not present in its collective-bargaining agreement, the arbitrator ruled against the company, and his decision was upheld by the court of appeals.[13] Although some arbitrators might rule differently, this case illustrates the advisability of gaining union support before introducing a policy not covered in the contract. Actually, it is good strategy to enlist union support from the beginning.

NLRB rules don't forbid employers from adopting a new nonsmoking policy if a union objects during collective bargaining. The employer is merely required to negotiate the issue in good faith. Because most employees are likely to support smoking restrictions, the likelihood of a strike over this issue is very small.

Employers should not bargain with a group of nonsmokers where a union is present if the nonsmokers are not the union's official bargaining unit. Doing so would invite charges from the union that the employer is violating NLRB regulations. On the other hand, where there is no union, an employer cannot merely ignore concerted action by employees. In this situation, according to Peck:

> Disciplinary action directed against a group of employees who are acting for their own protection, whether they are acting to get a no-smoking rule, or whether they are acting to protest a no-smoking rule involves the possibility that one would be found to have violated the National Labor Relations Act.[14]

## Other Legal Issues

An interesting twist to the legal aspect of banning smoking on the job is the Rehabilitation Act of 1973. Under this federal law, nonsmokers who can document a work handicap because of secondhand smoke will be entitled to protection under the law—reasonable accommodation by the employer, who may be required to create a smoke-free work area.

Smokers who can document a work handicap because of an addiction may also be entitled to reasonable accommodation. More commonly, though, it is nonsmokers who seek protection under the handicapped laws, in part because smokers may not want the stigma of "addict" attached to their employment. Federal Judge Donald Vorhees in Seattle upheld a nonsmoker's handicap claim under the Rehabilitation Act of 1973. He ruled that Lonnie Vickers, an employee of the Veterans Administration, was entitled to protection from secondhand smoke, which caused serious health problems.[15]

These mounting legal precedents can mean more costs to an employer, even if a nonsmoker leaves before the secondhand smoke actually produces illness. According to Lowenberg, "In Iowa and California unemployment benefits are now granted to nonsmoking workers who are forced to resign because of co-worker smoking and the refusal of employers to provide smoke-free work areas." This, of course, raises the employer's insurance premiums.

Section 5(a) of the Occupational Safety and Health Act of 1970 states:

> Each employer shall furnish to each of his (or her) employees employment and a place of employment which are free from recognized hazards that are causing or are likely to cause death or serious physical harm to his (or her) employees.

In several recent cases brought under the Occupational Safety and Health Administration (OSHA), the obligation of an employer to provide a safe workplace was deemed to be "unqualified and absolute . . . all preventable forms and instances of hazardous conduct must . . . be entirely excluded from the workplace."[16] And more than one nonsmoker who filed a worker's compensation claim for injury by smoke in the workplace has been awarded damages.[17-19] In light of cases like these, and others cited in Appendix 7, it is clear that employers risk more by *allowing* smoking at the workplace than by *banning* it!

No reasonable employer wants to abridge anyone's rights. But today, because the dangers of secondhand smoke have been widely recognized, the issue is correctly seen as whether employers have the right to subject two-thirds of the workforce to a hazardous substance that is removable. Public opinion and the courts are increasingly on the side of nonsmokers.

# References

1. "A Good Neighbor Policy: Controlled Smoking Areas and Health Department Attitudes," *Hawaii Medical Journal*, Jan. 1980.

2. "Smoking on the Job, the Controversy Heats Up." *Occupational Health and Safety*, Jan./Feb., 1979.

3. "Nichtrauchen am arbeitsplatz—eine umfrage unter rauchern und nichtrauchern im verkehrswesen," *Zeitschrift fur Erkrankugen der Atmungsorgange*, 1979.

4. "Passive Smoking at Work," *International Archives of Occupational and Environmental Health*, Spring 1980.

5. ". . . Office Staff Output," *Wall Street Journal*, April 17, 1981.

6. KIRO Newsline, June 12, 1983.

7. Project #82-63, Corporate Research Division, Human Resources Administration, Pacific Telephone, January 1983.

8. *Pacific Northwest Bell Employee Smoking Survey, 1983.*

9. "Do Employers Discount the Present Value of Smoking Employees?" unpublished paper by C. Patrick Fleenor and William L. Weis, Seattle University.

10. *Parodi vs. Merit Systems Protection Board*, 690 F. 2d 731 (1982).

11. *Smith vs. Western Electric Co.*, Missouri Court of Appeals, 643 S.W. 2d 10(1982).

12. Corporate Smoking Policies Conference, Bellevue, Washington, July 21, 1982, sponsored by the American Lung Association of Washington.

13. *Johns-Manville Sales Corporation vs. International Association of Machinists, Local Lodge 1609*, 621 F. 2d 756 (5th Cir. 1980).

14. Smoking OR Health in the Workplace conference, March 31, 1983, sponsored by the American Cancer Society, Washington Division.

15. *Vickers vs. Veterans Administration, et al.*, 549 F. Supp 85 (1982).

16. *National Realty & Co., Inc. vs. Occupational S. & H. R. Com'n*, 489 F.2d 1257(1973).

17. Workmen's Compensation Board, State of California, Case No. 76SF257-975.

18. Worker's Compensation Board, Hearings Division, State of Oregon, WCB Case No. 84-07248.

19. Fair Employment Commission, State of California, Case No. FEB 81-82 C8-0009ph.

# 10

# Reassurance for Executives

Employers considering new policies toward smoking can learn from the experiences of others. A 1979 survey by the National Interagency Council on Smoking and Health, found the following:

1. Half the companies responding had a current policy that either restricted or prohibited smoking in the workplace.

2. Fewer than 1 percent of the companies had ever calculated their financial costs from smoking.

3. Fewer than 1 percent once had a restrictive policy but had discontinued it.

4. While only 1.1 percent of the companies currently had incentive programs to encourage or assist workers to quit smoking, 33.2 percent wanted to develop or expand such programs. Among companies with more than 2,200 employees, 39.5 percent were interested.

5. Only 34.3 percent said they were not interested in developing a program on smoking and health for employees.[1]

Most employers on the threshold of a new smoking policy are looking more for reassurance than for statistics. Many fear the worst in the way of union challenges or employee rebellion. They worry, for example, that: (1) smokers—some of whom are important executives and managers—will quit their jobs; (2) smokers will become resentful and revengeful—and may even file suit; (3) employee morale will be eroded; or (4) union challenges will prevail and strike a blow to management control.

## A Successful Approach

Since 1973 the smoking policy at Boyd Coffee in Portland,

Oregon, has been simple. Employees who wish to smoke may do so only in the employee parking lot and only during morning and afternoon coffee breaks (15 minutes) and during lunch hour. Smoking is prohibited elsewhere on company premises—a rule that applies to everyone, even visitors and customers.

Boyd supplies restaurants and institutions in twelve western states with some 400 coffee and food products. The company has about 110 employees at its modern facility beside the Columbia River near Portland.

The company had been almost smoke-free since its beginning in 1900. Richard Boyd, personnel manager and secretary-treasurer, says that his grandfather simply believed smoking in food processing areas was inappropriate, so he restricted it to the lunchroom and entry lobby. In addition, the company's former building in downtown Portland was old and a possible fire hazard.

In January 1973 the company moved to its present location. Smoking was restricted to the lunchroom and entry foyer. On July 17, 1973, the following notice was posted on the employee bulletin board:

PLEASE . . .
NO SMOKING ON COMPANY PREMISES
Effective July 20, 1973

It has been established that smoking is injurious to health. It can also be unpleasant to those who do not smoke. It is our desire to furnish clean air to our employees and customers.

If you have been reading newspapers and periodicals in recent months, or watching TV, you will be aware of the findings in regard to pure air.

The State Legislature has banned smoking in public meetings, and there is an attempt being made to eliminate smoking in retail establishments by city ordinance.

If you feel, in spite of everything, you still must smoke, please do so *only* in your car.

[signed W. F. Rogers]

Richard Boyd described some of the events leading up to this posting. Since moving from the old downtown facility he had heard occasional complaints about the firm's restrictive smoking policy—complaints that the policy was not restrictive enough. The management team generally agreed that smoking on the job was unnecessary and was objectionable for sound business reasons. First, visitors to the company are often shown how coffee beans are processed. The aroma of fresh roasted beans permeates most of the facility, including the Little Red Wagon, a retail store for many of Boyd's products. But the lunchroom and entry foyer reeked of smoke, destroying the enticing fragrance. Second, evidence that passive smoking is harmful had begun to mount, and the Oregon Legislature had recently passed Senate Bill 508 restricting smoking to designated areas in public buildings, giving official recognition to the hazards imposed on nonsmokers.

Richard Boyd had hoped the matter would resolve itself without a potentially explosive confrontation between management and smoking employees. As personnel manager, he valued the respect and rapport he had developed with his employees. He didn't want that relationship endangered.

"I still remember the day," he recalled, "when Veda Younger, who was then secretary-treasurer of the corporation, stopped by my office with the news. Smoking was to be completely banned, and it would be my responsibility to implement the policy in whatever way I deemed most effective. I remarked that this thing could become a real can of worms, but I could tell that the decision represented a consensus among top management—and was final." Boyd was eager for dependable advice—but experts on implementing smoking restrictions were rare.

"I called Frank Wesson at Food Employers, whose background and knowledge in labor matters I valued. He suggested that we announce to the union that a policy to ban smoking from all company property was being contemplated. If the union responded within two weeks, then meet with union representatives. If the union failed to respond during those two weeks, then go ahead and issue the policy."

Boyd made the announcement and got no response. So he posted the notice on July 17, 1973, a Friday, to give employees a weekend to adjust to the new policy. Within a few days Teamsters Local 206 filed a grievance, saying the employees were not properly consulted. The issue went to a conference board, which deferred to binding arbitration. The arbitrator upheld the policy with one change: employees would be allowed to smoke in the parking lot, not just in their cars. The arbitrator thought smoking in cars would pose too much of a health risk because of the confined quarters and lack of ventilation!

"That was a concession we could live with," confided Boyd. "I suppose we experienced an initial two- or three-week period during which we detected the expected employee rumblings about the new policy. But after that there was a significant attitude adjustment, and we now receive very positive comments about the policy from our workforce. Job applicants generally react very favorably when we inform them of the policy." Today at Boyd Coffee the premises are noticeably cleaner throughout. Receptionists are especially cheerful and efficient. And the aroma of fresh roasted coffee beans is unpolluted by tobacco smoke.

Boyd Coffee has not calculated the exact savings from its smoking ban. But in 1980 its 110 employees totaled only 175 days of absence—a rate of 1.59 days per employee for the year. The *1981 Statistical Abstract of the United States* reported that men who smoked were absent from work an average of 8.5 days per year, and women who smoked were absent an average of 8.2 days per year.[2] Assuming that Boyd's workforce was half men and half women, its absenteeism rate in 1980 was only one-fifth that for all smokers!

Boyd's absenteeism rate was also below those for people who have never smoked (4.9 days per year for men and 6.1 days for women), which supports the claim that a significantly higher absenteeism rate can be expected among *nonsmokers* who work in tobacco smoke.

These data alone should help eliminate the anxiety managers often have over problems that won't arise. Boyd Coffee's experience also refutes the common excuse that "smoking bans are all right for other companies, but impossible at ours because we are unionized."

Few, if any, union agreements either state or imply a guarantee that workers can smoke on the job. A union can, of course, file a nuisance action. But, as noted in Chapter 9, if an employer and a union cannot agree on the development or the implementation of a policy, the employer is still free to adopt the policy if it has been negotiated in good faith.

Managers may still find themselves in an unwarranted tizzy. The personnel manager for a West Coast telephone company faced arbitration from an employee's grievance over a new smoking ban and was expecting turmoil to ensue from an unfavorable decision. The policy, which restricted smoking to a designated smoking lounge, was reasonable and had been tactfully implemented. The arbitrator ruled in favor of the company.

Even without unions, employers expect resistance—again without justification. Radar Electric's Warren McPherson related this experience:

> It was kind of a letdown. I managed to get myself thoroughly paranoid over all the troubles we were going to have with employees and customers when the smoking ban went into effect. Now here we are in the sixth year of our smoke-free policy and I'm still waiting for my first problem. I wish I could get back all those sleepless nights and give them to the people who warned me about all the serious problems I was going to have.

At a seminar on workplace smoking policies, Dennis Burns, president of Pro-Tec, a company that completely bans smoking, said simply to his astonished audience: "I'm sorry, we just haven't had any problems. What else can I say?"[3]

Oscar Austad, president of the Austad Company, reported similarly. "From the time we started in business in 1963, we have never permitted anyone to smoke on our premises. This includes every place in our building, no exceptions. We have never had a problem with this. I doubt that we have lost any business. We may even have gained some."

Regina Carlson, executive director of New Jersey Group Against Smoking Pollution (GASP) and author of an excellent monograph entitled *Smoking or Health in New Jersey*, has shared some of her correspondence with companies that restrict smoking. Without exception, the companies reported either no difficulty or minor difficulty implementing or enforcing smoking restrictions. For example, Merle Norman Cosmetics has enforced a no-smoking policy since 1977. Everyone adheres to the policy and it was accepted easily by the employees.

Falcon Safety Products, Inc., of Mountainside, New Jersey, which makes residential fire alarms and marine warning devices, prohibits any form of smoking on the premises except in a certain area in the warehouse. Company president Roy Thorpe assessed the policy this way:

> The ability to eat, work, and exist in a smoke-free atmosphere in fact has had a positive effect (in our opinion) on the work produced by our employees.
>
> The general reaction among nonsmokers was one of great satisfaction and the heavy smokers surprisingly had a very positive reaction to the program . . . I talked with each department manager to find out if the policy had had any negative effects on workers or productivity. The answer was a resounding "no" and, in fact, all indicated that because of the clean atmosphere workers appeared to be more alert and productive.

Group Health Cooperative of Puget Sound has 6,000 employees. Six months after the company went smoke-free, managers reported that there had been few violations, no enforcement problems, and no resignations resulting from the policy. Indeed, we have yet to hear of anyone who quit a job because of smoking restrictions.

Bob Gaston, managing editor of the Longview, Washington, *Daily News,* has described how employee complaints prompted him to do something about smoke in the newsroom. First he asked the smokers to be courteous, which accomplished nothing. He then instituted a complete smoking ban in the newsroom. To his surprise, he had no problems.

Several years ago Jeanne Weigum, personnel director for Immuno Nuclear Corporation in Stillwater, Minnesota, described her company's strict compliance with the Minnesota Clean Indoor Air Act:

> I did a survey of attitudes toward smoking restrictions. All line personnel participated . . . The survey results indicated 100 percent agreement with smoking restrictions in the workplace . . . A majority of people indicated that smoking restrictions at work were very important to them (as opposed to somewhat important or unimportant).
>
> It would be misleading to give the impression that the company is preoccupied with smoking and smokers. It is not. In fact, it is seldom an issue, and the regulations are a simple and acceptable fact of our workplace.

That message expresses what most smoke-free employers would want to convey to other employers who are apprehensive about restricting or prohibiting smoking: *Most employees favor a no-smoking policy because it helps them work better and feel better about working. Most smokers will support the policy because they, too, acknowledge nonsmokers' rights to clean air.* One or two smokers may object, but as we have seen, complications are rare and likely to be minor. Overall, the results are clearly favorable to both employer and employees.

Managers charged with implementing smoking restrictions may assume that employees, unions, and customers will create a wave of problems. Concern is appropriate; anxiety is not! As the next chapter shows, policies that are carefully designed and executed run into little or no trouble.

## References

1. *Smoking and the Workplace*, National Interagency Council on Smoking and Health, Business Survey, 1979.

2. *1981 Statistical Abstract of the United States*, Department of Commerce, Bureau of the Census, p. 123, Table No. 202.

3. Comments made at "Corporate Smoking Policies" seminar, Bellevue, Washington, July 21, 1982, sponsored by American Lung Association of Washington.

# 11

# Strategies for Going Smoke-free

Should a company decide to restrict smoking, there are several questions worth considering: Should a ban be total or partial? How should the transition be timed? How should policies be enforced? Should special incentives be offered? Let's look at these issues through the experiences of employers across the country who have gone smoke-free.

## Restrict or Ban?

Depending on the physical situation, some types of restrictions may not work at all—as demonstrated by the experience of Alaska Airlines in Seattle, Washington. Company officials decided to do something about drifting smoke after receiving complaints, especially from workers in the accounting department, where 30 to 50 people worked in one large room. The first attempt to accommodate both smokers and nonsmokers involved moving people around. That didn't work, so "smokeless" ashtrays were provided for the smokers. When that proved useless, the company spent thousands of dollars to alter its ventilation system. That didn't work either, so smoking was finally restricted to designated places.

There are sound reasons for a total ban. The ideal philosophy behind a no-smoking policy should be that the company is sincerely interested in its employees' health, working conditions, and general happiness. Maintaining designated smoking areas is a tacit message that smoking is acceptable if done in the "right" place or at the "right" time. That is not a healthy message. Permanent smoking areas can subvert the long-term goal of creating a smoke-free workplace. These areas may use valuable floor space, require costly ventilation, and tempt smokers to sneak extra breaks from work. But they are certainly better than having no restrictions at all.

## Fast or Slow Transition?

Because most people take time to adjust to changes, a gradual transition seems most logical. Having smokers involved in planning for the transition helps minimize any perception that the company is being arbitrary and dictatorial.

Tip Top Printing Company of Daytona Beach, Florida, followed both of these principles two years ago when it decided to prohibit smoking at its 40-employee office. It began with a ban from 8 to 9 a.m. one week, 8 to 10 a.m. the second week, and so on. The heaviest smoker, who was picked to chair the no-smoking drive, said he resented it at first but later was "really glad it happened."

In Minneapolis, Minnesota, the Park Nicollet Medical Center, the fifth largest multispecialty group practice in the nation, used an elaborate and carefully thought-out plan that began by designating one room in each building for smoking. Six months later, it started hiring only nonsmokers. Two years after the policy was implemented, smoking was not allowed on the grounds or in the buildings. The overall plan consisted of six major actions:

1. A survey of facilities to determine extent and variations of smoking.

2. A survey of employee attitudes (which showed 62 percent of 1,300 people favored a ban).

3. Establishment of incentives for healthy lifestyles, in part by reimbursing employees for completing such programs as smoking cessation and stress management.

4. A policy of hiring nonsmokers only.

5. A communications program to inform employees of the Minnesota Clean Indoor Air Act, results of the surveys, and the lifestyle programs.

6. Development of a policy options list. From this it was decided that an immediate ban would create unnecessary stress, that working gradually toward a ban several years hence would cause loss of momentum, and that the best time to ban smoking would be during the summer when employees could step outside to smoke.

Having employees vote can help them accept future changes. Things went smoothly after employees of the Kansas Department of Health and Environment voted 3 to 2 in favor of clearing the air. Smoking was then relegated to designated smoking areas, primarily lobbies, and is prohibited in hallways, offices, restrooms, and conference rooms.

Some organizations attempting to be "democratic" have let individual departments set policy under a general mandate. While this might make some sense because of individual variations in the number of smokers and the physical layout, severe problems can result. Some nonsmokers may not be afforded the protection they request. The story of Eldon Ball in Chapter 8 illustrates this problem. Inconsistencies cause additional problems when nonsmokers who want smoke-free working conditions don't get them. And permitting smoking only in private offices can create discontent among the smokers.

In one case, voting worked temporarily against the nonsmokers. At the Lehigh County mental health clinic in Bethlehem, Pennsylvania, all therapists have their own offices. Although smoking within these offices rarely caused smoke to drift outside and disturb nonsmokers, smoking by patients in the large waiting room created a serious problem throughout the clinic's hallways. Three years ago, employees (about half of whom smoked) voted against banning smoking in the waiting room—temporarily ending consideration of a ban for that area. During the next year, however, many clinic workers quit the habit (one after experiencing a collapsed lung) and those closest to the waiting room became disturbed enough about the high level of smoke that the group's feeling shifted and a ban was put into effect.

A few organizations have worked "backwards." Instead of giving priority to protecting nonsmokers at the worksite, they began by restricting smoking in lunchrooms, elevators, restrooms, stairwells, and meeting rooms—places where nonsmokers are *not* forced to gather. This was done to test the idea of nonsmoking areas or to begin gradually to phase in a complete ban. Although better than nothing, this approach puts nonsmoker protection last. Whether it prepares smokers for the major change they will eventually face is doubtful.

Notice of an impending ban seems to make good sense. Policies imposed without warning are likely to exacerbate employees' feelings—and justifiably so—that they have insufficient control over their lives. Advance notice can soften aggravation and reduce the tendency to resist.

At the Fred Hutchinson Cancer Research Center in Seattle, smokers were taken completely by surprise when a no-smoking policy went into effect. The Center had examined the problem through its personnel policy review committee. Upon the committee's recommendation, the Center's director issued a policy that "smoking is not acceptable within Center facilities." Some employees did not learn of the policy until two days after its effective date. Seven months later, a group of women was still gathering to smoke in an anteroom of a women's bathroom.

Other aspects of timing may have special relevance. Maupintour, in Lawrence, Kansas, went smoke-free when moving into a new building, giving its employees a good feeling of starting fresh. Weather conditions may also influence the beginning of smoking restrictions. Park Nicolett, discussed above, decided smoking restrictions would best begin during a period when smokers could step outside without freezing.

## Employee Involvement

Some organizations create special committees to develop and implement the policy through specific steps. This was done at Kessler-Ellis Products, Atlantic Highlands, New Jersey, where a management-employee committee on smoking devised a gradual transition. Thirty of the company's 85 employees smoked before the ban. Smoking hours were gradually reduced. At first the initial hour of work was declared smoke-free, then the last hour was added, then the hour immediately following lunch, and so on. "Smoking tickets" were rationed to smokers to spend for smoking breaks in the smoking area during no-smoking periods. The company proposed March 1983 as its first totally smoke-free month. But *smoking* employees insisted that the ban begin two months sooner!

Employee committees might also investigate and disseminate through classes or printed material supporting information for employees who want to stop smoking. This could include information on the available stop-smoking clinics, dietary support, lifestyle changes, and exercise, all of which could be integrated into the organization's transition toward the smoke-free policy.

Some of the more successfully orchestrated transitions have included a theme. The Ontario Ministry of Health launched Project Smokeless for 600 of its employees. Instead of no-smoking areas, smoke-free areas were designated and marked with attractive signs. On top of a purple background, an interesting geometrical design brings attention to the white letters below saying "This Is a Smoke-free Area," and below that, in smaller letters, "Help Make Project Smokeless a Success." To reinforce the theme, employees were given cardboard pencil-holders with "Help Make Project Smokeless a Success" on a red background. Notepads with Project Smokeless in purple ink were also made available. Communications from the smoking policy committee were made on stationery similar to the notepads.

## Policies Should Be Firm

When an organization commits itself to a smoke-free work environment, it should be prepared to enforce its smoking ban unconditionally. Good faith does not accommodate exceptions. For example, allowing supervisory personnel to smoke because they may work in enclosed cubicles promotes resentment among smokers who aren't allowed to smoke in open areas. The company should be prepared to discipline offenders because, if anyone is allowed to openly disregard a ban, it will quickly fail. Part of the Hutchinson Center's violations can be attributed to the lack of enforcement provisions in its policy.

After Boyd Coffee instituted its ban in 1973, the first employee found smoking in one of the restrooms was fired for violating the new rule. Merle Norman Cosmetics also threatens to fire instantly anybody who smokes in a restricted area. A company either has or does not have a smoking ban.

Firmness, of course, can involve some flexibility. Because severely addicted smokers may have a very difficult time adjusting to a smoke-free policy, a company may allow for occasional "slips." Group Health Cooperative of Puget Sound has such a policy. Violations of the smoking ban are treated in the same way as work-rule infractions—with progressively stronger disciplinary steps culminating, if necessary, with discharge.

## Incentives

Many organizations going smoke-free have added incentives to help smokers quit. The most common type is a bonus. The Flexcon Company, a specialty paper company in Spencer, Massachusetts, offered employees $30 a month to become or remain nonsmokers—and gave $15 a month to those who cut down on tobacco use. Cybertek Computer Products in Los Angeles offered those who had quit smoking for a year a $500 bonus.

Somewhat more innovative, Intermatic Inc., in Spring Grove, Illinois, opened an "I Quit Smoking" betting window. Smokers could bet up to $100 for double their money back at the end of the quit-smoking project. Another $1,000 was added to be distributed among the winners. Another type of approach would be to offer extra time off for not smoking.

## Nonsmokers Should Also Be Rewarded

Nonsmokers may resent it if those creating the smoke can collect a bonus for ending a bad habit, while nonsmokers get nothing for encouraging a change and for being more valuable employees to begin with. When nonsmokers can benefit from bonuses as well, an employer can generate more enthusiasm among all the workers for the policy.

Incentives to remain a nonsmoker are important for another reason: stopping temporarily is much easier than quitting for good. Bonuses that encourage a healthier lifestyle help provide long-term support and prevent relapses. In line with this philosophy, the George W. Dahl Company in Bristol, Rhode Island, paid employees (including existing nonsmokers) three dollars a week to refrain from smoking. Even better, Chuichi Yoshioka, a tool manufacturer in Kobe, Japan, paid nonsmoking employees $37 to $49 more a month in the beginning. With the added production from healthier employees and reduced cleaning costs, he was able to raise the bonus to $113 per month.

Another successful technique is the policy used by Radar Electric of combining sick leave and vacation time into "discretionary time off." Because sick days cut into days available for leisure activities, Radar's policy provides everyone with an immediate incentive to avoid illness. The only restriction in this time-off plan is that the absent employee's job must be covered by others. So discretionary time off promotes not only health but also cooperation and responsibility among employees. This policy eliminates the problem of employees calling in sick when they are not. And it simplifies the keeping of attendance records.

## Effects of Strong Leadership

Support from top management can make a tremendous difference in winning support from employees. One outstanding example is the program of Riviera Motors, Inc., near Portland, Oregon. A major distributor for Volkswagen, Porsche, and Audi products, the company employs approximately 650 workers. Its strategy for helping employees adjust to a smoke-free workplace was so effective that an actual smoking ban turned out to be unnecessary!

Riviera's "wellness program" dates back to 1977, when president Knute Qvale decided to make company headquarters in Hillsboro, Oregon, off limits to smoking. Thinking that an outright ban would be too painful for the nearly half of Riviera's workforce who smoked, company leaders began an aggressive campaign to help smokers kick the habit and refocus their interests on more healthful and wholesome activities.

The program included the following:

1. Smoking cessation classes sponsored by the Portland Adventist Hospital were held in company facilities. The company paid full cost for the eight 1-hour sessions. An hour of each session was on company time. Employees taking these classes were supplied with free fruit juice to ease the withdrawal symptoms.

2. Films were shown on wellness topics, some on company time.

3. Reduced membership rates were made available at a nearby racquetball club.

4. Exercise classes were held four days a week during the noon hour. They included aerobic dance, weightlifting, and yoga

(January through May, when the employees were less likely to exercise outdoors).

5. Literature on smoking cessation and health was distributed.

6. A bonus of $200 was offered to smokers who stopped for one year.

7. Volleyball games were held on Wednesday nights.

8. A summer softball league for women was developed.

9. Job applicants were told "up front" about the company's concerted effort to discourage smoking.

Riviera estimates that the smoking rate among its employees fell from 48 percent to 17 percent during the first four years of the program, and was 10 percent by March 1984. And it is clear that the ex-smokers have benefited from the continued emphasis on healthy lifestyles.

According to vice president James Thornburg, the company had been prepared to restrict smoking to certain areas and to stop hiring smokers. As it turned out, neither step was necessary because the percentage of smokers dropped so sharply.

Employee reaction? Purchasing agent Jean Hogan credits the company for helping both her and her husband break the smoking habit. (Spouses of Riviera employees are also eligible and are encouraged to attend the stop-smoking clinics.) Mrs. Hogan had smoked a pack a day, and her husband smoked 2-3 packs a day for thirteen years. "I feel very good about quitting and especially appreciate my employer's real concern for my health," she said. "I think most employees here feel very positive toward Riviera for supporting this program."

Executive secretary Betty Highhouse expressed similar feelings: "Like most heavy smokers, I wanted to quit, but I definitely needed the kind of help and support that Riviera provided. I really don't expect that from my employer, and when I get it, well, you feel very good about your company." Betty specifically commended Riviera's president for his leadership in promoting the company's interest in health.

Warranty adjustor Bob Grimm had smoked for forty-five years before quitting through the company's program. "I definitely needed my employer's help to do this—I wouldn't have even tried otherwise. They really give us support here—bending over backwards to help by providing fruit juices and other things when we really need them." Warehouseman Allan Vaught and inventory control specialist Alan Tolliver conveyed the same positive attitude, confirming that the health programs enhanced employee morale throughout the company. And Riviera's few remaining smokers are courteous and cooperative.

Riviera's support and educational system for the health of its employees is not a token gesture. The bright attitudes, improved lifestyles, and dedicated work performance all demonstrate employee appreciation.

With this laudable success, Riviera expanded its health program by building a fitness facility for its employees at its Hillsboro headquarters. The facility includes racquetball and volleyball courts, showers, lockers, a weight room, and a gymnasium. When the percentage of smoking employees drops to five percent, the company may take another look at its policy of hiring both nonsmokers and smokers. Remarkably, the organization is nearly smoke-free without the heavy hand of a formal policy.

## Overview

The best package an employer can offer is one that combines financial incentives, wellness advice, lifestyle changes, and on-site cessation classes with a strictly enforced policy that either prohibits smoking on company premises or permits it only in designated areas during employee break time. The designated areas should be isolated so that nonsmoking employees are never exposed to smoke. Common areas, such as lunchrooms and restrooms, should *never* be selected as smoking areas. In lunch-rooms, a smoking area should be maintained *only* if nonsmokers will not get a wisp of smoke in the nonsmoking section. Designated smoking areas should be for smokers only.

Although a few smokers may attempt to organize coordinated opposition to the new policy, an employer is likely to overestimate the extent to which this will occur. The experiences of many organizations indicate that few smokers will rebel and that objections will die quickly if violations are dealt with firmly.

At the time a company announces its intention to ban or restrict smoking on the job, it should *seriously* consider a simultaneous announcement that henceforth only job applicants who are nonsmokers will be considered. This policy is the best long-term insurance for a smoke-free workplace.

It is far easier to implement and enforce clear, unequivocal work rules than to wrestle with exceptions and conditions that degrade health and morale. And the benefits of a complete policy are clearly superior to those with exceptions. Whatever the variations, it should be obvious that clean air is good for both people and profits. As far as we know, no employer who has established a smoke-free policy has ever abandoned it!

# 12

# Smokers Need Not Apply

All of us make choices. We choose a red car instead of a blue one. We choose Channel 5 because it has better programs. Choosing can be considered a form of discrimination—of selecting what we want over what we don't. Usually we try to choose what we think is best for ourselves. Sometimes we pick what we think is best for others.

Organizations use a similar process when selecting employees. From a stack of 1,000 job applications, for example, only two are needed to fill two open positions. Which applicants are the best? Developing job-related skills would be a waste of time if employers drew resumes from a hat instead of selecting people whose characteristics appear best for the job.

Such characteristics as race, sex, age, and national origin are beyond individual control and are usually irrelevant to job performance. Discrimination on the basis of such attributes is illegal and, in the opinion of most Americans, is immoral as well. But when it comes to education, most people have control over their choices. Employers have always considered education important for some jobs and have "discriminated" among applicants accordingly. The same reasoning is being applied to smoking as more employers realize that hiring only nonsmokers makes economic sense.

If an organization commits itself to becoming smoke-free, it certainly makes sense to stop hiring smokers—especially heavily addicted ones. Hiring nicotine addicts to work in a smoke-free work environment would be like hiring claustrophobics to work in a submarine. Does it make sense to continue to hire smokers into an environment that may unsettle their nerves? Both logic and good faith seem to say no. Restricting future hiring to nonsmokers (or smokers determined to quit) will ensure in the long run a smoke-free organization without complications.

## The Application Process

Radar Electric, which hires only nonsmokers, makes this policy clear during the application process. Its application form asks applicants whether they smoke, and those interviewed are informed that Radar Electric is a nonsmoking company. Smokers who are not hired are thus spared needless self-doubts, confusion, and loss of self-esteem. On the other hand, some companies that don't hire smokers do so without informing applicants.

Another good reason to ask whether job applicants smoke is to ensure their understanding and acceptance of the nonsmoking policy. Smokers who perjure themselves to get jobs will undoubtedly learn to live with the policy or quietly pursue employment elsewhere. They are unlikely to smoke while at work and will certainly not lead a rebellion against the smoke-free policy.

How can an employer tell whether a person smokes? First, most smokers, and certainly heavy smokers, cannot function long in a smoke-free environment—and they know it. Most smokers won't lie on an application form unless they are reasonably sure they can get through the workday without smoking. If they actually can, their addiction is minimal. So an applicant who professes to be a nonsmoker is probably either a nonsmoker or a light smoker who can abstain during work hours (and will suffer much less damage to health and be absent less than a heavy smoker). This person may also be interested in breaking his addiction and may welcome the support of a smoke-free environment. The important principle here is to hire only people who describe themselves as "nonsmoker" on a written application form. When this answer is backed up with a written policy for the employee's personnel file, the employee has recognized and accepted the organization's smoke-free policy.

It is important to note that imposing a hiring ban is far easier than imposing a ban on workplace smoking. The latter involves a transition that may be difficult for smokers because of withdrawal symptoms and changes in customs and employment conditions.

Hiring only nonsmokers can be an immediate and painless step toward a smoke-free environment and will back up the smoke-free policy. Current laws support such a hiring practice because they are based on *achieved* characteristics, including smoking. Consider, for example, the achieved characteristics commonly used to compare job applicants:

1. prior work experience
2. level of education
3. job skills
4. grade-point average
5. employer references
6. willingness to accept company policies (e.g., dress code, working hours, conduct at job stations, etc.)
7. test scores
8. willingness to relocate
9. willingness to work overtime
10. personal grooming.

It could be argued that some of these attributes are arbitrary and unrelated to job performance. The courts today are sometimes sympathetic to employees who claim damage because of arbitrary discrimination unrelated to performance. But in light of the discomfort imposed on other employees and the higher costs associated with smoking at the workplace, no one will succeed in arguing that smoking behavior is not job-related.

## Legal Facts

The "employment-at-will" doctrine, upon which labor laws in the United States are based, frees employers to establish hiring and firing criteria subject only to specific legislative exceptions. This means that private employers may discriminate against job applicants on *any* basis except those specifically prohibited by statute. Those prohibited now include race, color, religion, sex, and national origin (Title VII of the Civil Rights Act of 1964), union membership (National Labor Relations Act), and certain physical handicaps (Rehabilitation Act of 1973).

Public employers (government agencies) are bound further by the Equal Protection Clause of the Fourteenth Amendment, which has been interpreted to require that job qualifications be subjected to a judicial test of "strict scrutiny" (where the qualifications are based on a "suspect" class) or "rational basis" (where the qualifications do not affect a "suspect" class). "Suspect" classes include minority races and religions—but not smokers. Hence a public employer would merely need to show a rational justification for its nonsmokers-only hiring policy, such as a desire to minimize costs and protect employees from hazardous substances during working hours.

## Can Smokers Be Fired?

Hiring only nonsmokers is one thing, but what about firing employees who smoke? "Mr. Smith," an independent personnel specialist who wished to remain anonymous, was asked by *Wall Street Journal* reporter Jennifer Hull in 1982 whether it would be discriminatory to discharge an employee for being a smoker. In reply, Smith cited the case of *Metropolitan Dade County vs. Wolfe*, 274 So. 2d, 584(1973), cert. denied, 414 US 1116(1974), which sustained the discharge of an employee who was 53 pounds overweight and therefore "more likely to become disabled." The discharge was based on the precedent set by *Griggs vs. Duke Power Co.*, 401 US 424(1971) that any "job-related" factor is proper. Foreseeable disability was held to be job-related.

Noting that smoking is far more deleterious than mere overweight, Smith also addressed the employees' right to smoke in general: "The Supreme Court has made clear that there is no right to smoke, when it upheld a total ban on cigarettes in the case of *Austin vs. Tennessee*, 179 US 343(1900)."

He went on to cite *Spencer vs. Toussaint*, 408 F. Supp. 1067 (E.D. Mich. 1976), which ruled that excluding people with a history of mental illness is not unreasonable. Tobacco dependence is classified as a mental illness by both the International Classification of Disease and the Diagnostic and Statistical Manual of Mental Disorders.

Smith concluded:

> Employers have a right, indeed a duty, to protect safety; smoking poses a foreseeable danger to the smoker, to property, and to other persons . . . Discharges for smoking do not involve prohibited discrimination. An employer has the right to control what is done on duty time. An employer has an obligation, however, to have a "clean" case and not use smoking as a pretext.

On April 15, 1982, in an article titled "Burned Up Bosses Snuff Out Prospects of Jobs for Smokers," Ms. Hull reported basically the same message from the federal Equal Employment Opportunity Commission: "Isn't it illegal to discriminate against smokers? No . . . unless the result is discrimination on the basis of national origin, race, religion, sex or age. If, for example, an employer hired men who smoked but not women, there could be a problem, says an EEOC spokesman."

A few months later, at a conference on corporate smoking policies held in Bellevue, Washington, Cornelius J. Peck, professor of law at the University of Washington, was asked whether hiring only nonsmokers was legal. Peck replied that, under the employment-at-will doctrine, employers may discriminate for any reason not specifically forbidden by law. He then drew an analogy to sexual harassment: "Until Title VII of the Civil Rights Act established that sexual harassment was illegal, there used to be many employers who imposed upon women working for them by demanding sexual favors . . . What was the remedy? The answer used to be none." Peck added that he didn't see anything that would make it against public policy or business justification for an employer to say, "We don't employ people here who smoke."

On the "Today Show" of August 2, 1982, where Warren McPherson of Radar Electric outlined how he had made his company smoke-free, interviewer Jane Pauley remarked twice that hiring only nonsmokers was presumably legal—and posed that statement as a question to the Tobacco Institute's Tom Howard. Mr. Howard agreed without hesitation that such a policy was legal!

That understanding is not widely shared among management personnel. Employers still confuse ascribed characteristics like race, age and sex with achieved attributes like smoking. Though it's a credit to our equal employment consciousness that most of us shrink at the mention of "discrimination," we should remember that choosing on the basis of achieved qualities is at the core of our free-enterprise system.

Those inclined to weep for the plight of unemployed smokers should put away their handkerchiefs. Most employers will continue to hire smokers. But for job-seekers looking for a smoke-free workplace, the opportunities will increase. The world will be more livable when smokers work for employers who don't mind polluted environments and nonsmokers work for employers committed to health and good management practices.

# 13

# Breaking the Tobacco Habit

It's the puff of glamour. The smile of sin. The grit of pain. All rolled into one. But the ball and chain of tobacco habituation drags more than 500,000 Americans a year into the grave.[1] U.S. Surgeon General C. Everett Koop calls nicotine "the most addictive drug in this country."[2] Some may argue that nicotine addiction is mild. But the U.S. Department of Health and Human Services sees it this way:

> Four drugs stand out among all drugs and substances of abuse. Fewer than 500,000 persons use heroin, but with other opiates it exacts a terrifying toll of crime and social disarray. Alcohol affects 10 million problem drinkers and their families and accounts for half our automobile fatalities. Marijuana is the most widely used *illicit* drug. Tobacco, in the form of cigarettes, is smoked by 56 million Americans. It causes more illness and death than all the other drugs.[3]

Despite these frightening facts, it's not easy to stop smoking, because smoking is habit-forming as well as addicting. Dr. Jack S. Henningfield, behavioral biologist at The Johns Hopkins School of Medicine and chief of the Human Performance Laboratory at the National Institute of Drug Abuse, notes:

> Just 10 or 15 years ago, if you wanted to quit cigarettes, people told you to "just quit," to use willpower . . . to put your cigarettes in a different drawer. Now we realize that this is absurd—like telling an alcoholic to put his bottle in a different drawer or telling an opiate addict to put his needle in a different part of the house.

Henningfield and colleagues have identified 13 ways that nicotine addiction resembles opiate addiction;[4] and he cites a one-year study that found that abusers of alcohol, opiates, and tobacco relapse at about the same rate.[5] At the end of one year, 80 percent of the tobacco and narcotic addicts and 77 percent of the alcoholics had returned to drug use.

Bruce Hanson, an ex-smoker who is co-director of Dependency Intervention, a counseling agency in Berkeley, California, stresses that many smokers perceive the world differently and don't learn to solve many of life's problems. In the habituated smoker's mind, "Things go wrong because there was no cigarette . . . and will be better with a cigarette," Hanson observes. "When you're addictively involved you're kind of on psychic hold in some ways . . . Once you use a drug in an addictive manner, it's with you for life. You're always in danger of going back."

For Frankie Smith, the motivation for quitting was the threat of dying. In 1976, at age 39, she lay in intensive care so near to death from emphysema caused by smoking that last rites were administered. "My physician had written on my chart, 'This woman will not live out the year. If she should lose consciousness, do not attempt to resuscitate her.'" Her indignation at this instruction sparked her back to life.

"Addiction is hating yourself," says Ms. Smith, who has since developed Kicking It, a smoking-cessation course in San Francisco, based on the book of the same title by David Geisinger. Along with exercise and imagery, a major emphasis in her course is the politics of smoking. "Smoking is a political act. The government and corporations are generating revenue at the expense of people's lives," Smith declares. "Unless you show smokers how to stay aware of what's going on out there in society, their chances of staying off cigarettes are not good."

## How America Got Hooked

Most Americans born before the mid-1930s were weaned on a steady diet of tobacco glamour. *Life* magazine's full-page testimonial advertisements featured filmland's superstars of the 30s, 40s and 50s. Among those who died prematurely from smoking-related afflictions were:

- Gary Cooper, who told why he switched to Lucky Strikes, after years of selling his testimonials to Chesterfields
- John Wayne and Dick Powell, who extolled the supposed manliness of Camels
- Joan Crawford, who touted Camels for women who also wanted to be tough
- Bing Crosby and Ed Sullivan, who pushed Chesterfields
- Robert Taylor, for Lucky Strikes.

It didn't matter whether screen actors and actresses were recognizable through the smoke of a wartime cabaret—so long as their names were on the marquee. Cigarette smokers also brought America the sounds of the Big Bands. Aficionados of that era still enjoy Chesterfield's recorded broadcasts of Glenn Miller from the Cafe Rouge and the Glenn Island Casino.

White-coated "physicians" recommended smoking by explaining why most doctors smoked Brand X. Gary Cooper testified that it was good for the throat and voice. Even Ronald Reagan got on the bandwagon for Chesterfields. Above all, the ads implied, heroism, romance, and sex just couldn't happen without the magic wands of success: cigarettes. The message was loud and clear: Everyone who was anyone—even Franklin Delano Roosevelt himself—smoked.[6]

Much has been written about the winners and losers of World War II. The biggest winners, however, were J. P. Lorillard, Philip Morris, American Brands, and R. J. Reynolds. Free cigarettes for military personnel meant that millions of newly addicted customers would share America's post-war economic spoils with the tobacco industry.

Many of the war's biggest losers languish today in Veterans Administration hospitals across the country. According to Seattle lung specialist Rolf Holle, half the patients are there because of smoking. Others have already died. Their enemy, the cigarette, remains undefeated—an enemy that kills more Americans each decade than have been killed in all of our country's wars going back to the American Revolution.

Today our whole nation provides a gigantic display of the carnage of smoking. But chastising those born before 1940 for getting hooked is like ridiculing polio victims for contracting a disabling disease before discovery of an effective vaccine. Americans were lured during an era that glorified tobacco before its hazards were clear.

## Today's Realities

Cigarette companies have spent huge amounts of money to promote social acceptance of smoking. In 1983—the latest year for which figures are available—the advertising tab was a record $2.7 billion. But young people today, who have grown up with overwhelming evidence of smoking's damage to health, don't look sophisticated, mature, glamorous, or sexy when they smoke. They look stupid. They look stupid to the 35-year-old executive who runs marathons for fun. They look stupid to 60-year-old smokers who believe, rightly or wrongly, that they would never have started smoking had they known of its devastating effects. How sad that tobacco companies can undermine people's ability to know and trust their own bodies. The violent cough with the first puff is a danger signal. But billboards and magazine ads deliberately avoid this connection and encourage people to put common sense aside.

Bright young people today see smoking classmates as *losers—*

immature children in adult carcasses incapable of exercising good judgment regardless of medical facts or bodily warnings. But to look upon an older generation with similar scorn and ridicule is wrong. That generation was lured to addiction without a mountain of medical evidence to direct it down a healthier path.

## Factors in Quitting

During the past 30 years, medical evidence and social pressure have persuaded millions of smokers to break the shackles of addiction. Would-be quitters must cope with three possible problems. Those who are physically addicted may have to suffer through a withdrawal period that can last for days and may include symptoms of nervousness, irritability, difficulty in concentrating, and a severe craving for cigarettes. Those for whom smoking is ingrained as a habit must find ways to break the habit. And those who rely on smoking as a means of "relaxing" or reducing tension must find other ways of dealing with their tension. The most significant factor in stopping is probably the strength of the individual's desire to stop.

The benefits of quitting are clear. No matter how long one has smoked, it is still beneficial to stop. During the first day after quitting, the heart and lungs begin to repair the damage caused by inhaled smoke. Smoker's cough usually disappears within a few weeks, energy and endurance may increase, and taste and smell may return for foods that haven't been enjoyed for years. After 10 years, the risk of dying from heart and blood vessel disease goes down to the level of nonsmokers. The risk of lung cancer decreases, and so do the incidence of respiratory infections and lung tissue destruction responsible for emphysema. A study of more than 10 million ex-smokers found that the death rate 10 years after quitting approached that of people who had never smoked.

Most smokers who quit do so on on their own, but stop-smoking programs can play an important role too. Some of the best known are the American Lung Association's Freedom From Smoking clinics, the American Cancer Society's FreshStart program, the Seventh-day Adventist Church's Five-Day Plan, and SmokEnders. All attempt to help participants identify the reasons they smoke and to develop better behavior patterns to replace the smoking behavior. Schick Clinics, available on the West Coast, utilize an adverse conditioning technique that supposedly makes the smoker sick of smoking. Many communities offer other programs through local hospitals and health departments. Some programs set up emotional support through "buddy systems" and follow-up visits. Nonprofit programs usually cost from $5 to $40, while commercial ones can cost several hundred dollars.[7]

Smokers with a strong addictive component to their habit may be helped with nicotine chewing gum. Available with a doctor's prescription, the gum can control withdrawal symptoms while the individual works on behavioral change. To work best, it should be combined with counseling by an experienced physician. A comprehensive book dealing with this subject, *How to Stop Smoking—Permanently*, by Walter S. Ross, can be obtained for $5.20 from Little Brown and Company, Distribution Center, 200 West St., Boston, MA 02254.

Smoking cessation programs have their limitations. One study found that only four percent of smoking employees offered a substantially reduced fee to take the SmokEnders treatment enrolled even though more than half of the smokers surveyed wanted to quit and indicated willingness to attend a clinic.[8] Long-term cessation rates for those who do attend are usually between 20 and 35 percent.[9-11]

But when smoking at work is banned, success rates are greater! New England Deaconess Hospital reduced its employee smoking rate by 39 percent in just 20 months through a comprehensive policy restricting smoking to designated areas in the hospital. Rita Addison, former executive director of the Clean Indoor Air Educational Foundation, helped design and implement the hospital's policy. A major selling point for her services, tailored to an industry focused on health, is that the *most* effective agent for smoking cessation is a working and living environment that is smoke-free. "Little effort was given to directly promoting smoking cessation," says Ms. Addison. "In fact, only two of the 380 employees who quit did so through an on-site cessation program. Therefore, the significant smoking decrease came about primarily as a direct influence of policy modification."

Radar Electric's Warren McPherson agrees. "When we began phasing in our smoking ban at Radar, half of my 100 employees smoked. Now, four years later, there are only two 'closet' smokers left, and no one has left the company because of the policy." Radar's policy did not include the use of an on-site smoking-cessation program. Smoking restrictions are clearly a powerful inducement to break the tobacco habit.

The Austad Company has not permitted smoking since its inception in 1963. Later only nonsmokers were hired. In 1985, only 10 of its 225 employees were smokers. Company president Oscar Austad says that nonsmokers throughout the United States have been willing to move to Sioux Falls so they can work in Austad's smoke-free environment.

# References

1. Ravenholt, R. T., "Addiction Mortality in the United States, 1980: Tobacco, Alcohol and Other Substances," *Population and Development Review* 10:697-724, 1984.

2. *Listen* magazine, July 1983, p. 13.

3. *Why People Smoke Cigarettes*, U.S. Department of Health and Human Services, Public Health Service Publication No. (PHS) 83-50195, 1983.

4. Henningfield, J., et al., "Human Dependence on Tobacco and Opiods: Common Factors," in T. Thompson and C. E. Johanson, eds., *Behavioral Pharmacology of Human Drug Dependence*, National Institute on Drug Abuse Research Monograph 37, DDHS Pub. No. (ADM) 81-1137, 1981, pp. 210-234.

5. Hunt, W. A., et al., "Relapse Rates in Addiction Programs," *Journal of Clinical Psychology* 27:455, 1971.

6. For additional examples of tobacco advertising, see Blum, A., "What the Surgeon-General Didn't Tell Us," in Barrett, S., ed., *The Health Robbers*, George F. Stickley Co., Philadelphia, 1980.

7. Kiefhaber, A., and Goldbeck, W., *Smoking: A Challenge to Worksite Health Management*, p. 18, (background paper for the National Conference on Smoking OR Health, New York City, March 18-20, 1981.)

8. Kanzler, M., Zeidenberg, P., and Jaffe, J. H., "Response of Medical Personnel to an On-site Smoking Cessation Program," *Journal of Clinical Psychology* 32:670-674, 1976.

9. Evans, D., and Lane, D.S., "Long-term Outcome of Smoking Cessation Workshops," *American Journal of Public Health* 70:725; 1980.

10. Schwartz, J. L., and Rider, G., *Review and Evaluation of Smoking Control Methods*, DHEW, Public Health Service, 1978.

11. Shiffman, S. M., "The Tobacco Withdrawal Syndrome," in Krasnegor, N. A., ed., *Cigarette Smoking as a Dependence Process*, DHEW, 1979.

# 14

# Your Guide to Speaking Up

Nonsmokers who want clean air at work may face a serious challenge to their psychic endurance. The Tobacco Institute would like us to think that simple "courtesy" can solve this pollution problem. But many nonsmokers, as courteous as their stamina will allow, have found that getting clean air involves a struggle. They've had to battle against inconsiderate co-workers, management doubletalk, and endless frustration. While some nonsmokers have prevailed through employee pressure, others have wound up in court.

One of the latter is Marie Lee, a financial assistance worker at the Attleboro Community Service Area Office in Attleboro, Massachusetts. In 1979, after she complained, her director issued a memorandum to all staff that: "Effective immediately, there will be no smoking in the open areas of the office. The lunchroom is the designated area for smoking." But smokers ignored the memorandum. (So much for courtesy.) So in January 1983, Ms. Lee sued for relief [Massachusetts Superior Court Department, Civil Action No. 15385]. Two years later, the Commonwealth of Massachusetts agreed in an out-of-court settlement to provide smoke-free work areas at the Attleboro office.

## Preliminary Steps

Before immersing yourself in an all-out effort to achieve clean air, ask some pertinent questions: "Do I really like this job? Am I using skills I want to use? Aside from the smoke, would I like to be doing something more challenging? Is it time for a promotion, and an office of my own? Have I been thinking of other kinds of work? Or of moving?"

Many employers seem to value their machinery more than their employees and are reluctant to change to a no-smoking policy. So from the start, develop plans in case you: (1) get fired; (2) must go to court; (3) must apply for workman's compensation or unemployment benefits; (4) are harassed by management or co-workers; or (5) get no support from the union. Learn what each contingency requires so you'll be ready if it occurs.

## Document Everything!

Each time something related to smoking happens at work, record the date, time, place, what happened, who was involved, and what was said (as close to verbatim as possible). For example, if your sinuses ache after someone lights up, write that down. Also take the time to make a detailed floor plan of your work area, including ventilation ducts. Keep your notes at home, and don't talk about them to co-workers. This documentation is essential, for without it, you cannot prove your case.

You will need a statement from a medical doctor about your reactions to secondhand smoke. It is also wise to locate a sympathetic lawyer in case you need one. Local nonsmokers' rights associations may refer you to one who is suitable. Action on Smoking and Health (see Appendix 10) may offer legal advice to your lawyer. Additional tips can be found in *Smoke in the Workplace: An Action Manual for Non-Smokers* (1985), by Martin Dewey, available for $6 plus $2 postage from the Non-Smokers' Rights Association, Suite 201, 455 Spadina Avenue, Toronto, Ontario M5S 2G8, Canada.

## Create a Base of Support

Fighting for clean air is easier if you're not alone. So determine how people around you feel about the smoke. Observe their reactions throughout the workday. Do they cough more when it is smoky? Do they back away when someone lights up? Observations like these can be used to encourage nonsmokers to talk about their feelings. For example, if a person coughs during smoky meetings, make a sympathetic comment like, "I noticed you were coughing. The smoke bothers me, too." This could easily lead to further sharing of feelings and serious discussion that will accomplish two things: (1) confirm and reinforce another person's dislike of smoke, and (2) help that person to express dissatisfaction and complain to others about the situation.

If your company has an existing committee of employees or managers to deal specifically with the problem of smoke in the workplace, join it. If not, perhaps you can get one started.

Creating support outside of work is also important. After a hard day, it may be good to share your frustrations and anger with another person. Better yet, find several friends with whom you can talk, so that you don't risk overburdening one individual. This tactic has other advantages, too. Should your attempts to get clean air become drawn out, you may need help with research, fundraising, and media exposure. The more supporters you have, the greater your chances of success.

## Join an Established Group

Nonsmokers' rights groups are committed to the idea that nonsmokers should not be *forced* to breathe the smoke of others. Located throughout the country, they have gradually been persuading smokers that courtesy requires determining—before lighting up—whether anyone else minds. These groups are also responsible for the growing number of state and local laws to protect nonsmokers in public and at work. So join one of the groups listed in Appendix 10 of this book and support it with labor, money, or both.

## Stockpile Supportive Literature

Whether the initial suggestion to clear the air is written or verbal (which should be reinforced by a written suggestion to document your efforts), provide information on the consequences of second-hand smoke, the advantages of eliminating it, and suggestions on how to accomplish this. Many groups, such as the American Heart Association, the American Lung Association, and the American Cancer Society, have pamphlets available. Additional facts can be used from this book, and copies of scientific reports may be pertinent. Be sure to include your personal observations. If you are affected physically, cite your discomfort in detail. This information should be attached to memoranda, grievances, and letters when you make your preferences known to superiors and union officials. Always keep copies for yourself.

## Follow Protocol

When filing grievances or complaints, pursue each option in the correct order. A procedural error could be used against you. Instead of addressing the issue, management (or later a court of law) could say instead, "This wasn't submitted properly, therefore we have no obligation to consider it." Your written appeals should indicate the procedural basis of your complaint.

Company rules should be researched for any provision that provides for disciplinary action for committing an act detrimental to the health or safety of a fellow employee. If your organization has such a rule, press for its application to tobacco smoke. For employees who work for a temporary help agency, the usual procedure is to complain to the employer, not the client.

## Exhaust Administrative Remedies

In her landmark book, *How To Protect Your Health At Work* [Environmental Improvement Associates, Salem, N.J., 1976], Donna M. Shimp, who sued New Jersey Bell for a smoke-free workplace, advises afflicted workers to appeal to all local, county, state and federal agencies, even though they may not be able to do anything. It is important to document that you have "tried everything," should you end up in court. All requests should be in writing and sent certified mail, return receipt requested. Ms. Shimp suggests that at least these organizations be solicited:

> Environmental Protection Agency
> U.S. Public Health Service
> State Department of Labor
> State and County Health Departments
> Office of the Public Advocate
> Occupational Safety and Health Administration
> National Labor Relations Board.

## Talk to Smokers

It is a good idea to communicate with each smoker as diplomatically as possible about your goals. Even though sympathy from smokers may not be forthcoming, informal notice can help reduce tension and enable them to prepare for what could be an important change. Talking shows respect and may help reduce the bitterness that can accompany a forced change, even one that is health promoting.

Another way to reach smokers is through advocacy. In an informal conversation, mention to a smoker, preferably an empathetic one who might be willing to set an example, that your nonsmoking colleague is having a hard time because of the smoke. Efforts of this type can help smokers become interested in the problem. The more smokers you can get to campaign in your behalf, the better.

## Circulate a Petition

A petition can demonstrate the sentiment of nonsmokers, who are likely to be in the majority. Before passing it around, try to obtain your supervisor's permission so you are protected if anyone objects afterwards. If a union is present, the employer will not be able to bargain with those signing the petition, but it will still demonstrate sentiment. Smokers as well as nonsmokers should be given an opportunity to sign and indicate whether or not they favor a smoking restriction.

## Avoid Air Purifiers

Air purifiers—provided they are powerful enough—can screen out pollen and particulate matter that pollutes the indoor air. Room-sized models are rarely effective, and so-called "negative ion generators" don't work at all. However, use of an air purifying device may provide an excuse for continuing a policy that allows smoking.

# Introduce a Stockholder's Resolution

In many companies, stockholders have the right to submit resolutions that are voted on at stockholders' meetings. This method helped prod the huge Boeing Company, based in Seattle, to begin steps toward a smoke-free workplace.

In November 1983, the secretary-treasurer of FANS (Fresh Air for Nonsmokers, Washington State) advised Boeing that he and FANS, both holders of Boeing stock, would introduce the following resolution at the next stockholders' meeting:

> Resolved: That the stockholders of the Boeing Company . . . hereby request the Board of Directors to take the steps necessary to control the burning of tobacco products in all areas inhabited by nonsmokers. The discomforts and health hazards produced by burning tobacco are well documented and result in reduced productivity, increased absenteeism, maintenance, health and life insurance costs, etc.

Copies of the letter were sent to local, national and financial media and to local attorneys. Meanwhile, many nonsmoking Boeing employees filed complaints with management about tobacco smoke. After securing a lawyer's opinion that Boeing could be held liable for keeping cigarette machines on its premises, FANS solicited the names of deceased Boeing employees who had smoked in order to investigate possible lawsuits by their survivors. In April 1984, just two days before the stockholders' meeting, Boeing announced its intention to become smoke-free.

The Securities and Exchange Commission (SEC) requires those who submit resolutions to have $1,000 worth of stock. If a local clean-air organization is willing to be a co-sponsor, stock can be donated or temporarily transferred to it for that purpose.

Once the resolution is drafted, submit it by certified mail, return receipt requested, with copies to the SEC and appropriate media. FANS also suggests sending copies to attorneys sympathetic to the clean air movement. Be sure to list the "cc" recipients on the original. If the company refuses to put the resolution into the proxy material, complain to the SEC.

In the meantime, ask the company how to buy a list of their stockholders' names and addresses and the procedure for nominating candidates for their Board of Directors in advance of the meeting. From the list (which may include members of the clean-air organization), solicit proxies for voting power. At the meeting, introduce the resolution and nominate an appropriate person to serve on the Board of Directors.

The rules for submitting resolutions and nominations vary. Some companies require six months or more advance notice, so check the requirements carefully.

## Assert Yourself

As you begin to make yourself heard, you may be harassed by co-workers or even management. In the face of such opposition, it's good to be assertive but not aggressive. This requires tact and ingenuity, not brawn. The fundamental principle is to treat smokers as people, just as you would expect them to treat you. Yet, because they feel attacked and insulted, many nonsmokers act rudely toward smokers.

What folly! Jumping on smokers only makes them defensive and nervous. To cope, they may even smoke more. Both sides then lose. A far better method is diplomacy that avoids friction completely. A polite request puts people at ease, allowing—rather than trying to force—them to comply. If you aren't certain how to phrase your requests, consult your colleagues or friends who smoke. Explain the situation and ask their advice.

If you are a timid sort of person, not used to confronting people, learning to speak in public may help you. The Toast

Masters and Mistresses are designed to do just that. Books like *How to Develop Self-Confidence and Influence People by Public Speaking*, by Dale Carnegie [Pocket Books, 1977], and *Don't Say Yes When You Want to Say No*, by Herbert Fensterheirn and Jean Baer [Dell, 1975], are worth reading.

Mastering your own feelings is also important. Are you berating yourself for being in an uncomfortable situation, as though it was your fault? If so, instead of blaming yourself, channel your energy to create better options. Work toward clean air or toward finding a different job. Remember, you have a perfect right to object to tobacco smoke and to decide whether you want to suffer as a result of someone else's addiction. So instead of saying, "I guess I deserve to work in smoke" or "There is nothing I can do about it," think of constructive steps you can take. This can keep you working toward your goal, generate self-trust, and increase your ability to influence your work environment.

## Communicate Tactfully

Tension between smokers and nonsmokers can be minimized by using language that avoids personal attacks. After all, the primary problem is the *smoke*, not the smokers. We therefore suggest that you refer to the smoke, not the smokers; to the health of *all* employees, not just that of nonsmokers; and to the goal of clearing the air, not just eliminating visible smoke.

Just as the words and the way you deliver them are important, so are the physical expressions of your message. Show a clenched fist with wild eyes and you may provoke hostility before you open your mouth. Show openness and reassurance and others will listen to you. The more comfortable you make others, the more they will accommodate you. Let your eyes trace their eyelids. Try to breathe easily (even in smoke). Place yourself at a friendly distance. When standing, face to one side of the person you address. Facing squarely may make others feel attacked and react defensively.

## Signs Should Be Pleasant

Perhaps you have seen the bright fluorescent orange signs with black letters that say: "NO SMOKING PLEASE." They get everybody's attention, don't they? And they are rather ugly, too. Such signs have their place, but would you want one on your desk or outside your door? Do you want everyone to think you lack aesthetic taste? Unsightly signs work against the clean-air cause. Instead of building an image of health and beauty, they may convey the idea that clean-air crusaders have no sensitivity. But a well designed sign will convey the desirability of clean air.

The quickest way to understand the power of aesthetics and visual imagery is to look at the cigarette ads that depict attractive outdoor scenes. How effective do you think those ads would be if they showed sickly emphysema patients in a smoky room? Take a tip from the tobacco companies: choose your images carefully. Use signs and posters that reflect the health and beauty of clean air, not your disgust for smoke.

## Try to Avoid Arguments

Asking a smoker to hold back for a while often leads to an endless debate about rights. Once you get into one of these arguments, you will have a hard time getting out. Usually, nobody wins, so why expend the energy?

Your goal should be to obtain the cooperation of smokers to *abstain* in your presence. That should require not an argument but assertiveness and the ability to stick to your issue. Some retorts from smokers that can lead you astray are: "Who are you to speak with authority?" "What difference does it make, the air is polluted anyway?" "Why don't you go outside if you want clean air?" and so forth. Your reply should be: "I understand how you might disagree with me, but the smoke still bothers me. Will you please not smoke here?"

If a smoker says you are oversensitive, you can reply, "It doesn't make any difference what label you want to put on me, my eyes burn just the same, and I have to deal with that."

## Press for Smoke-free Meetings

The best way to deal with smoking during meetings is to prevent it. The effort needed to accomplish this depends on your attitude and on the attitudes of the smokers. Some may be willing to try a smoke-free meeting; others may hesitate or be downright rude. Keep in mind that you are not trying to get anyone to stop smoking, just to abstain during the meeting. Talking with each smoker individually and casually ahead of time may get you several commitments to try a smoke-free arrangement. Then ask the chairman to place the issue on the agenda for discussion.

If you are chairman and have the right to forbid smoking in the meeting room, a good way to announce your decree is: "I have been asked by many of you to request that there be no smoking in this room." Posting a sign often helps because everybody can see immediately there is to be no smoking. Don't forget to remove the ashtrays.

Outside of work, when registering for meetings or seminars, include a note to remind the sponsors that many nonsmokers, including yourself, will appreciate a smoke-free learning environment. The more people who do this, the more smoke-free meetings there will be.

## Smoke-free Travel

For those whose work includes travel, asking for clean air shouldn't stop outside the office door. For a while, the Interstate Commerce Commission (ICC) allowed smoking in the rear third of buses that cross state lines. Now, smoking is permitted in the last four rows only. This restriction includes the driver.

Despite the regulation, smokers may light up even if they aren't in the last four rows. Asking a smoker to comply may spare the driver the tension. But if the situation requires that you go forward to the driver, do so. Ask the driver to make an announcement to remind passengers of the smoking regulations. The driver has the right to remove any passengers who don't comply. However, most drivers won't do this unless something bizarre happens.

Complaints against drivers who smoke while driving or who refuse to enforce smoking regulations should be made as soon as possible—even during a break or layover. Union contracts sometimes forbid disciplinary action unless the report is made within a few days of the incident. Among the details you should note are the coach number, time, date, last city of departure, destination, and the driver's name if posted. Letters should be sent by certified mail, return receipt requested, to: Secretary, Interstate Commerce Commission, Washington, DC 20423. Send a copy to the bus company and ask for a reply. Ask for part of your money back if you feel strongly about the inconvenience.

Some states completely prohibit smoking on interstate and intrastate buses. If your state lacks such a sensible law, write to your state legislator and to the ICC and ask that a total ban on smoking be enacted.

On local buses, smoking is usually forbidden. Yet some smokers who are ignorant or rebellious will light up whether the bus is empty or full. People like this may believe they have a right to smoke anywhere at any time. Asserting yourself directly to this kind of smoker, as you may be tempted to do, can be dangerous.

We know of one nonsmoker, for example, who asked a smoker seated behind him to stop. The smoker returned a cold stare, said nothing, and snuffed the cigarette only after the nonsmoker went forward and appealed to the driver. Twenty minutes later, the nonsmoker got off the bus and walked two blocks to the post office. The smoker followed behind and tried to provoke a fight. The moral: Not all smokers are rational; so size up the situation carefully before you act.

In the United States, Amtrak is the largest and most frequently used rail service for travelers. Compared to other methods of commercial transportation, Amtrak is heaven for nonsmokers. Smoking is forbidden in nearly all coaches and in the dining car. With more nudging, smoking and no-smoking restrooms could probably be designated.

If you see someone smoking where it is prohibited, you could politely remind that person. Seeking the aid of the conductor or porter can also be effective. Complaints and recommendations about no-smoking policies should be directed to: Consumer Relations, Amtrak, 400 N. Capitol NW, Washington, DC 20001. Be precise in your letter and offer a suggestion or two.

Under current Department of Transportation (DOT) regulations, anyone who arrives on time and requests a seat in the no-smoking section of a commercial airplane shall be provided one, even if the airline must convert a smoking row into a nonsmoking one. Failure to comply subjects the carrier to possible enforcement action by DOT, which can order a fine. Unfortunately, any money collected goes to DOT rather than the aggrieved party. If passengers could collect damages (which is now possible, but only through a lawsuit), there might be more complaints and fewer violations.

Complaints about smoking on airlines should be directed to: Secretary, Department of Transportation, 400 Seventh St., S.W.,

Washington, DC 20590. In your complaint, provide as many details as possible: flight number, date, time, where the alleged infraction took place, and anything else that you may think pertinent. Comments about the smoking situation, perhaps urging a total ban on smoking, should be sent to the same office.

Other DOT rules affect nonsmokers. Action on Smoking and Health (see Appendix 10 for its address) will send you a free wallet-size card with these rules so you can always carry them with you.

Don't forget that the time others have taken to write their complaints and comments has changed all-smoking planes to ones with no-smoking sections. This is proof that individual action—on the road or at work—can make a difference.

A few hotels and motels offer rooms for nonsmokers. If there is no sign to this effect, ask for a nonsmoker's room. The clerk may be instructed to offer one to people who ask. Even if none is available, asking will help create a demand that will eventually have to be met.

When making reservations by phone or mail, drop in a line that says you would appreciate a room that has been occupied only by nonsmokers. If you find yourself in a room with a bad tobacco odor, go to the front desk, explain the discomfort, and ask for another room. Appearing in person instead of calling on the phone will emphasize your preference. A sensitive clerk will note that you went out of your way to make the request and will try to please you.

Before checking out, fill out any forms or cards that ask you to rate the accommodations. Give the establishment a nice comment if they assigned you to a room with no cigarette odor. Such praise will encourage them to continue their policy. If you were bothered by the previous tenant's smoke, include the comment in an otherwise good report.

## Smoke-free Eating

No matter how fancy the food, smoke can make you think you are eating out of a lunch box in a chemical factory. Many restaurateurs and patrons seem to think that smoking is an essential part of the meal. (Restaurants sell cigarettes, but do not provide clean air or gas masks.) Always ask for the no-smoking section whether one is apparent or not. Public pressure will eventually lead more restaurants to clear their air.

Should you find yourself sitting next to a table of smokers, you could request another table. (Apologize to the staff if you disturbed the setting before you decide.) Or ask a waiter or the manager to ask the smokers to stop. Should you decide to approach the smokers yourself, you must be gracious. When you speak, lead with an acknowledgment that they are probably enjoying themselves. "I presume the food is good? I have a rather difficult time enjoying mine when I am around smoke." Many smokers are reasonable and will refrain until you have finished your meal. If you are direct and confident, and politely look the person in the eyes, you will have a good chance of getting your request.

Business meals may be more complicated because of the possibility of offending a potential client or business partner. You must remember, though, that smokers need to be considerate too, and that refraining is likely to bother the smokers less than their smoke will bother you. If clean air is important to you, when you arrive at the restaurant ask the group if a no-smoking section is agreeable to all. Even if someone objects and you end up in a smoky part of the restaurant, comments of this type will remind smokers that drifting smoke annoys people.

Whatever the situation—at work or on the road—remember that your action may make the difference between breathing smoke or clean air. When you can ask for clean air without being defensive or apologetic, you are helping to clear the air for everyone.

# 15

# A Model Timetable for Change

Until recently, most people did not perceive smoking on the job as a problem. Most nonsmokers accepted tobacco smoke as a normal part of living and kept gripes about it to themselves. Smokers, too, kept their lips tight—around cigarettes. "It was our God-given right to smoke anywhere, anytime," observed Darrell Douglas, a former smoker who is administrator of Oregon's workers' compensation program. "It didn't mean a thing that others suffered."

Today, employee complaints and legislative mandates are changing this situation. Consequently, employers may be looking for ideas on how to move from a smoke-as-you-please workplace to one that is smoke-free. This fictional account of how Clean Company's 115 employees made the change incorporates the best ideas from the experiences of companies that have done so already.

*January 3*: President Leed announces through a memorandum to all employees that one of his company's goals for the year is to become smoke-free. Leed cites smoking's hazards to both smokers and nonsmokers, the legal obligation Clean Company has to provide a safe and healthy place to work, and the economic and morale benefits other organizations have obtained from a smoke-free workplace. The memorandum invites participation on a 15-member Clean Indoor Air Committee (CIAC), composed of smoking and nonsmoking employees and union leaders, to develop a specific plan of action. Employees will be kept informed of the policy's development.

*January 10*: President Leed chooses the CIAC and convenes a meeting two days later. The committee decides to survey the employees and the facilities (in part to determine if space is available for smoking areas) to help determine the proper steps.

*January 17*: A survey similar to the one in Appendix 1 is sent to all employees. A subcommittee of CIAC tallies the results for presentation to the full committee on January 24. CIAC members inventory company facilities, looking for rooms that might be used temporarily as designated smoking areas.

*January 25*: Analyzing the results of the survey, CIAC decides that Clean Company can be smoke-free by July 1. The committee also decides that nonsmokers should be afforded protection as soon as possible. It therefore recommends to President Leed that designated smoking areas be created as soon as possible, that smokers should be given time to prepare for a total ban, that smokers be given help in becoming nonsmokers, and that incentives be provided to reward the nonsmoking lifestyle. CIAC also adopts a "Be Smoke-free" theme. Special "Be Smoke-free" stationery will be printed to convey information on the project and no-smoking signs will feature the same slogan.

*January 27*: President Leed and CIAC issue a joint memorandum that restriction of smoking to designated areas will begin within the next six weeks. A smoking cessation program will be offered on site during work hours the same week the restrictions begin.

*February 4*: After examining several stop-smoking programs, CIAC selects the Seventh-day Adventist's Five-Day Stop Smoking Plan for its in-house program. The committee also makes provisions for smokers to enroll in any other reputable clinic. Half of the fee will be reimbursed after six months of being a nonsmoker. A buddy system for smokers trying to quit will also be instituted. The designated smoking areas are chosen after considering their convenience and air flow. A few will require special ventilation to prevent smoke from reentering the workplace. President Leed promises to take care of this problem.

*February 11*: Leed reports to the committee that the necessary ventilation for the designated smoking areas will be in place within a week. The committee and Leed agree that designated smoking areas will become effective within two weeks. The long-term goal of Clean Company is to be totally smoke-free. On the back of Clean Company's business cards will be printed: "Clean Company is promoting the health of its employees by becoming a totally smoke-free organization. Won't you please help us Be Smoke-free by refraining from smoking inside?" It is understood that, if a customer enters with a lit cigarette, an employee will present the card in a friendly manner, with little comment and minimum eye contact. The cards will also be displayed on the counters. CIAC and President Leed also agree on penalties for violating smoking restrictions. A first offense will draw a warning; a second offense, a suspension, and a third offense, discharge.

*February 12*: President Leed and CIAC issue a memorandum informing employees that beginning February 28, smoking will be restricted to designated areas listed in the memorandum. The memorandum also announces the enforcement provisions. The smoking cessation program is announced with a request that smokers sign up early if they are interested. All employees are solicited to become "buddies" for those trying to quit, with pairing to be done by a CIAC subcommittee.

*February 18*: CIAC and President Leed determine the financial incentives for nonsmoking employees. Wages of nonsmokers will be raised 75 per hour, beginning with the first paycheck after a full month of designated smoking areas. To further promote the health of all employees and to encourage smokers to quit smoking, sick leave and vacation time will be combined into discretionary time off.

*February 25*: A memorandum is issued to remind employees that beginning in three days smoking will be permitted only in the designated areas. The memo announces that wages will go up for nonsmokers in six weeks. And in two months, the discretionary time-off policy will take effect. The memorandum also reminds smokers that the smoking cessation program begins Feb-

ruary 28 and encourages smokers who are trying to quit to feel
free to talk with their supporting partners.

*February 28*: At the beginning of the day, no-smoking signs
are put up throughout the building. Ashtrays are removed from
the work areas and cafeteria and moved into the designated
smoking areas. Ashtrays are also placed next to the entrances so
customers can leave their cigarettes behind without littering the
premises. The Five-Day Stop Smoking Plan begins.

*March 5*: CIAC meets and learns that no problems have
arisen so far. The committee also discusses the idea of promoting
the health of all employees through weekly lunch-hour lectures
on nutrition, exercise, safety, and related subjects. A subcommittee is set up to solicit ideas for topics and to secure expert
speakers.

*March 19*: CIAC reviews progress to date and approves the
agenda of health talks proposed by the subcommittee. The subcommittee will send out an announcement of these talks within
the next week. The CIAC decides that conducting a survey about
a month before the total smoking ban begins would be wise in
order to spot any last-minute problems. The survey would also
serve as a reminder of the forthcoming ban. A six-month follow-
up survey is also planned.

*June 4*: CIAC finds that the survey results show overwhelm-
ing support for the policy so far and for the July 1 date to Be
Smoke-free. The results of the survey are summarized in a notice
to employees advising them that on July 1 smoking will be
prohibited inside the building. A special Be Smoke-free Day
celebration is planned for that day. Cheese, crackers, and vege-
tables will be provided by Clean Company to employees and
customers. News releases will be sent to local and national media
inviting them to report the transition. One of the rooms formerly
designated as a smoking area will become a permanent health
resource center for employees, with health-promoting literature
provided by the Clean Company. The rest of the designated

smoking areas will be converted into lounges.

*July 1*: The employees are in good spirits. Many customers are impressed with the celebration and say they will talk to their own employers about becoming smoke-free. Local TV and newspaper reporters drop by for additional information. President Leed tells them things have gone smoothly.

*August 5*: The CIAC meets, reviews the progress, and decides that the transition has gone so well that no more meetings are necessary until the six-month survey results are in.

*December 10*: CIAC meets, makes final changes on the survey, reviews the results, and reports back to employees that a substantial number of the employees approve the program and wouldn't think of returning to a smoke-as-you-please environment. Employees like the new smell of Clean Company and its cleanliness. Some employees comment that they are no longer reluctant to wear their nicer clothes to work out of fear that the dirt and grime might cause damage. Other employees say that work has become more pleasant and that they no longer think about leaving just to escape from the smoke. The increase in wages is appreciated by everybody. The best news is that 40 percent of those who smoked before the new policy began have kicked the habit. The prospects for Clean Company to become entirely smoke-free look good!

# 16

# Political Action

Each time a new law is passed to protect nonsmokers at work or in public places, the tobacco industry cries, "Government intrusion!" It doesn't seem to care that being forced to breathe drifting tobacco smoke is intrusive and unfair. The simple way for management to thwart this legislative trend would be to voluntarily provide smoke-free areas wherever nonsmokers need them. Until this is done, nonsmokers have but one choice: to seek protective legislation.

## Why Tobacco Companies Are Opposed

The tobacco industry, which appears to be concerned only about profits, is terrified by anything that discourages smoking in public. It knows well that as smoking becomes less socially acceptable, the number of smokers will drop. And it also knows that social or legal restrictions will decrease the number of cigarettes smoked per person. An article in the November 26, 1984, *Advertising Age* estimates that, if 60 to 70 million smokers give up one cigarette a day because of restrictive legislation or social pressure, 22 billion fewer cigarettes a year will be smoked.

That number translates to about $1 *billion* in retail sales, and each additional drop of one cigarette per day per smoker represents another *billion* dollar loss. Little wonder the tobacco industry is willing to fight tooth and nail and to spend large amounts of money to oppose local or state legislation that would restrict smoking.

In San Francisco, for example, the industry worked hard against Proposition P—voted on November 8, 1983—which upheld a previously passed ordinance to protect nonsmokers at work. Through a front organization called Concerned Citizens Against Government Intrusion, the industry spent $1.1 million to oppose the ordinance, while San Franciscans for Local Control spent $133,000 to oppose it. (The latter group, a coalition led by Californians for Nonsmokers' Rights, included the California division of the American Cancer Society, various units of the American Lung Association, and other groups.) Despite this financial disparity, the nonsmokers won by a tiny margin.

During the campaign, tobacco forces claimed that the ordinance was an invasion of people's privacy and would require exorbitant use of city funds for enforcement. Claims were also made that putting up partitions between smoking and nonsmoking areas would cost businesses millions of dollars. The tobacco industry's message apparently scared some people, including the editors of the *San Francisco Examiner*. On November 6, 1983, an editorial entitled "Smoke and Nannyism" urged San Franciscans to vote against Proposition P. The editors argued that a law to restrict smoking would create a "lot of grief and misuse of city resources" and that the ordinance was "a recipe for antagonism and frustration in the workplace, and there is nothing healthy about that result." Could the tobacco industry have been so eloquent?

Contrary to these dire predictions, newspapers around the country began to report success as early as two months after the ordinance went into effect (March 1, 1984). By May 20, 1985, 14 months later, there had been only 144 complaints, a few more than the number received *each month* on other problems. Bruce Tsutsui, the Environmental Health Inspector in the Department of Public Health who responds to complaints, says he spends about one day a week on them.

After a complaint is filed, Tsutsui calls the complainant and ascertains the situation—kind of workplace, whether a written policy exists, and what kind of violations are alleged. He then makes an on-site visit. If the complaint appears valid, he discusses with the owner or manager how to resolve it and relays proposed solutions to the complainant. After agreement is reached in this manner, Tsutsui telephones the employer and sends a written notice with 15 days to comply. A follow-up visit is then made to determine compliance. Only rarely have there been instances of noncompliance after the notice has been sent—and these violations have been from smoking employees disregarding the employer's policy. Tsutsui advises employers that they, not the city, are responsible for enforcing the new policy. Only once has noncompliance led to an administrative hearing, the next step after continuing violations. A $500 fine can be imposed after an administrative ruling is violated, but no fine has been needed so far.

During this process the identity of the complaining nonsmoker is not revealed to the employer, a policy that protects the nonsmoker from harassment or discharge for seeking clean air. Only under subpoena can the name be disclosed. Tsutsui says that smokers who become upset by the new restrictions calm down within a few weeks.

But for many nonsmokers not afforded protection by law or employer wisdom, smoke remains a big problem. And the reason they are actively seeking protection similar to San Francisco's is illustrated in the response to a nonsmoker's plea by former Seattle City Light superintendent Joe Recchi:

> . . . It has been almost eight months since your suggestion was submitted and I apologize for this delayed response . . . Until there are specific federal, state or city regulations prohibiting smoking, City Light will continue to allow smoking in the work area.

That memorandum was written on March 22, 1982. Today the nonsmoking employees of Seattle City Light are still working in smoke. With such blatant disregard for employee health and well-being, it's no mystery why many nonsmokers are willing to walk a mile—or more—for legislation to clear their air.

## Success Requires Hard Work

Provisions to protect nonsmokers at work are often combined with others to restrict smoking in public places to form a smoking-pollution control bill. Californians for Nonsmokers' Rights has drafted a model bill for this purpose (see Appendix 9). The usual procedure in seeking passage is to find one or two sympathetic representatives to introduce the bill and to get as many co-sponsors as possible.

Legislative victories require good planning and dedicated effort. In addition to making private presentations to each legislator, successful activists swamp elected officials with supporting letters and phone calls.

Testimony at public hearings may also be crucial to demonstrate support in front of the media.[1] Possible speakers include local representatives of a nonsmokers' rights organization, the American Lung Association, the American Heart Association, the American Cancer Society, a public health department, and a school of public health or medicine. Concerned individuals might include asthmatics, owners of smoke-free businesses, management consultants, physicians, and other health-care professionals. The more people who show up—if only to help fill up the hearing room—the better. But those who wish to give testimony should coordinate their statements to ensure that repetitious ones do not crowd out more important comments. If speakers talk in the order of sign-up, arriving early may push opponents to the end of the list.

Who are the opponents? Tobacco lobbyists are virtually certain to be there. If the proposed legislation includes smoking in public places, expect a restaurant association representative. If the workplace is included, expect a business association representative. Regarding the battle to restrict smoking in public places and city-owned worksites in Overland Park, Kansas, organizer Gloria Cohen has reported:

> Tobacco brings in lots of people to appear at hearings and council meetings. They all claim the "rights" are being taken away from them. "Too much government," they say . . . I was very aware of the opposition during our presentations: shuffling, squirming, whispering, coughing, anything to be distracting.

## Events in Overland Park

Based on her experience in generating support from the 80,000 citizens of Overland Park, a suburb of Kansas City, Cohen suggests:

> Put notices of what you're trying to do on bulletin boards of your office, grocery store, church, in local papers, church and synagogue newsletters, even door-to-door flyers. There are people out there with similar ideas, but they just don't know how to reach each other. Wear nonsmoking buttons. People constantly stop to ask about them, and a dialogue gets started on clean air and possibly another supporter is aboard. Petitions carry weight. We set up a table at a shopping center, collected names, and gave out Lung Association materials at one point.

A portion of Report No. 84-21, August 17, 1984, prepared for the mayor and city council by Larry Flatt, director of community development, sums up the work of Cohen and her colleagues:

As a result of concerns expressed by citizens concerning the
health hazards of tobacco smoke in public places, the
Community Development Committee [of the city council]
commissioned a citizen task force to develop recommendations to address these concerns. The Smoke Task Force
met over a four-month period and recommended ordinance
refinements and an educational program to address this
matter.

The task force recommended:

● Use of the slogan "Breathe Free" in all elements promoting
the Overland Park Clean Indoor Air Act (OPCIAA).

● Printing of 10,000 bumper stickers with the slogan "Breathe
Free" to help kick off the campaign.

● Proclamation of a "Smoke Awareness Week."

● A promotion program directed at restaurants, including:
table signs; a 6-star award for restaurants offering nonsmoking
areas; and designation of restaurants with nonsmoking areas in
promotional material produced by the Overland Park Convention
and Tourism Bureau.

● A poster contest in Overland Park elementary schools.

● Development of an audiovisual presentation and speakers'
bureau for promoting smoke-free environments in workplaces
and public gathering places.

● Provision by the city of standard signs in at least two sizes
for posting in "no smoking" areas.

● Adoption of resolutions 1820 and 1822 setting forth the
city's policy on smoking and nonsmoking areas in public buildings, in city employee work areas, and at other city activities and
events.

● A survey conducted by junior and senior high school
students to determine citizen awareness of OPCIAA.

The resolutions were approved by the city council on September 17, 1984, and the ordinance was passed on October 8, 1984.
The 12-member Smoke Task Force included two council mem-

bers, two restaurant owners, one bank employee, one shopping-mall manager, the director and administrative secretary of community development, one citizen at large, and one representative each from the Committee for Nonsmoking in Public Places and the American Lung Association. The twelfth member, Terry Frakes, of Austin, Texas, represented the Tobacco Institute. According to Cohen, Frakes had shown up in nearby Olathe, Kansas, when its council met to consider an ordinance similar to that of Overland Park.

## The Fort Collins Referendum

Overland Park's experience seems easy compared to the effort expended by the citizens of Fort Collins, Colorado. On April 3, 1984, the city council voted unanimously to adopt Ordinance 23 to prohibit smoking in any enclosed indoor area open to the public or serving as a place of work, except in designated smoking areas. At the workplace, fully enclosed offices and rooms assigned exclusively to smokers were to be identified, but employers had to provide smoke-free work areas upon request. But GROAN—Growing Resentment Over Anti-smoking Noises—challenged the ordinance with enough petition signatures by the end of May to require voter approval in the November election.

Support for the ordinance was led by Coloradoans for Clean Indoor Air (CCIA). In June the group generated letters to the editor and guest editorials, and also supplied writers with information on secondhand smoke and the benefits of smoking restrictions. News releases were sent to the media about the latest research demonstrating the health hazards of secondhand smoke. On public access cable TV, CCIA participated in debates with the opposition.

CCIA began its more aggressive "official" campaign with a news conference on September 13, 1984. Speakers focused on three main issues: the freedom Ordinance 23 would provide both smokers and nonsmokers, the adverse health effects of passive smoking, and tobacco industry intervention into local affairs— even though tobacco industry involvement was not yet obvious.

In mid-September, the city council authorized $10,000 to explain Ordinance 23 to the citizens. (The amount actually spent was only $725 plus $1,062 for staff time.) In mid-October, CCIA started a door-to-door campaign, providing residents with a copy of *Smoke Signals*, a 4-page newspaper with the headline, "If you don't want your child to be a secondhand smoker . . ." Immediately below the headline was a picture of a young boy looking at his father through a cloud of tobacco smoke. The paper included articles explaining indoor air pollution and Ordinance 23, letters from fifth graders about smoking in restaurants, a guest editorial from a city council member, a summary of the health hazards of secondhand smoke, and an article, "Who Do *You* Believe," informing Fort Collins residents they could tell the Tobacco Industry to "Butt Out."

The tobacco industry tipped its hand just 17 days before the election when the Committee Against 23, funded by the Tobacco Institute, registered with the city. Serving as treasurer was Mike Biggs, an athletic trainer from Colorado State University. By Monday, the media were replete with the Committee's opinion. Radio ads like this ran as frequently as twice an hour:

> Hello, this is Mary Pier, co-owner of the Wine Cellar Restaurant. I'd like to share my feelings with you regarding an issue we'll all be voting on November 6. My opposition to ordinance 23 involves the negative feelings I have about political interference in private business. I myself don't smoke. I'm not necessarily for or against smoking as a health issue. I think it's an individual choice. As a business owner of this community, the thing I resent most is the government telling me how to regulate a business that I put my heart, my soul and my money into. Eight months ago

the Wine Cellar voluntarily designated a nonsmoking area as a courtesy to our customers. Business owners don't need government telling them what to do. The bottom line dictates that. As a citizen and local businesswoman I urge you to join us in keeping government out of our lives. Please vote against ordinance 23. It goes too far. Paid for by the Committee Against 23. Mike Biggs, Treasurer.

Later came shorter radio ads that began:

Here are the facts:
Ordinance 23 means government interference in our personal lives.
Ordinance 23 means 90 days in jail and a $300 fine.
Ordinance 23 means more taxes for a bigger tax burden.
Ordinance 23 means higher operating costs for businesses.
Ordinance 23 means you pay more for goods and services.
Ordinance 23 means Fort Collins citizens no longer have a choice.

The Committee Against 23 also mailed registered voters a postcard with similar claims. These assertions so outraged city officials that, at a news conference on November 1, Mayor Gerry Horak demanded the Committee retract the statements about increased taxes and about jail terms and fines for violators and business owners. The cost of enforcement, he said, would be $147 to $1,000 per year. Penalties, as with other misdemeanors, would be up to the court and rarely would involve the maximum allowed by law. And the ordinance didn't require anyone to report violations.

Ready for the tobacco industry media blitz, CCIA demanded and got equal time on radio through the Federal Communications

Commission's Fairness Doctrine. They had already prepared public service announcements, two of which were:

> There's an election coming up and there's a lot of strange talk coming down. Of course, when the facts are against you, all you can do is try to hide the truth. That's what the tobacco industry is trying to do. The tobacco industry wants to be left free to poison your air and abuse your health. That's the only issue, how much poison you will be forced to breathe. It's time to tell the tobacco industry the party is over. Vote yes on Ordinance 23. You can turn out the lights on the tobacco industry. Thank you. Sponsored by Coloradoans for Clean Indoor Air.

> There's a debate in Fort Collins, a debate over freedom. Some people don't understand what freedom is all about. Freedom is about fairness and that's why Ordinance 23 is fair to smoker and nonsmoker alike. Freedom is about responsibility and that's what Ordinance 23 asks everyone to be. Be responsible. Be fair. Be free in the great American tradition. Vote yes on November 6th. Vote yes for Ordinance 23 for the freedom to breathe free, responsibly, and fairly. This message was sponsored by Coloradoans for Clean Indoor Air.

CCIA also swung into action by placing a series of newspaper ads prepared by communications consultant Eric Lucas. One showed a man waving goodbye as he gets into a Cadillac. In his other hand is a "Tobacco Industry" briefcase with dollar bills sticking out. The ad read:

> Here today, gone tomorrow.
> Wasn't it nice of the tobacco industry to drop into town, tell us a few fibs about Ordinance 23, and take off?
> They told you, for instance, that tobacco smoke is about as innocent as mother's milk. Why regulate smoking

in public?

They told you violators of the ordinance would have to go to jail for 90 days and pay a $300 fine. Too extreme, they said.

The truth is that tobacco smoke is definitely hazardous to smokers and non-smokers alike. Tobacco smoke is the leading preventable cause of death in the United States!

And violations of the ordinance are punishable by the same penalties as any other city ordinance. Ninety days and $300 is the maximum penalty, and is only handed down in the most extreme cases—and in the past 30 years has *never* been given a first time offender.

You'll never see Mr. Tobacco Industry again, but the rest of us will live with the results of Tuesday's election. Will we vote for clean air or not?

Vote for 23. Freedom for both sides.

The American Lung Association of Colorado jumped into the media flurry, too, by mailing to every registered voter a brochure entitled "Let's Clear the Air!!!" The section on "Creating a Healthier Work Environment" cited lost workdays and lost profits from smoking, in addition to legal decisions favoring nonsmokers in the workplace.

On November 6, Fort Collins residents approved Ordinance 23 by a vote of 21,075 to 12,012. "By the time the tobacco industry showed up," says CCIA chairperson Connie Acott, "I believe a lot of minds had been made up. Because the foundation for their intrusion had already been laid down, the public did not appreciate their propaganda."

The tobacco industry, which outspent CCIA $35,000 to $10,000, was apparently bewildered by the defeat. So it funded an anonymous telephone survey to discern reasons for the vote. The Colorado-Wyoming Restaurant Association, which had also opposed Ordinance 23, fired off a letter to Larimer County members the day after the election that illustrates the opponents' surprise:

"Though we lost—and we still don't know exactly what happened and why . . . If you have any indication about why we lost so badly we would appreciate hearing about your findings."

Impressed with CCIA and Colorado GASP (Group Against Smoking Pollution)—which along with Californians for Non-smokers' Rights had provided assistance), the letter reiterates the tobacco industry fears: "Naturally our great fear is that the GASP group will get their 'steamrollers out of the barn' and approach many city councils throughout the State." And that's exactly what's happening, not only in Colorado but in many other communities across the United States.

## Campaign Tips

Connie Acott, now with GASP of Northern Colorado, advises campaigners to keep these general principles in mind:

● *Organize early* and get as many different groups as possible involved. Professional health organizations like the lung, heart, and cancer groups, physicians, and nurses groups are obvious candidates. Others to keep in mind include hospitals, public relations and advertising agencies, labor unions, and environmental and educational groups. The "Associations" category in the *Yellow Pages* is a good resource.

● *Develop background information* that can be provided to letter writers, speakers, fund raisers, and the media.

● *Raise funds* through selective mail solicitations and personal appeals. Donation cans with attractive labels can be placed in stores and other locations to collect money and provide exposure.

● *Set up a telephone tree* to rally people when quick support is needed.

● *Set up a speakers bureau* whose speakers can explain the benefits of the legislation to groups and the media.

- *Develop a good public image* by creating attractive, well-written materials; attacking the tobacco industry rather than local opponents or smokers, who are also the victims of the tobacco industry; stressing the legislative benefits to smokers and non-smokers; and sticking with the issue to avoid arguments.
- *Keep messages simple* so voters don't become confused.
- *Monitor all information going to the media* so leaders and the media committee can coordinate media strategies.
- *Use the media as much as possible* by encouraging people to write letters and guest editorials; creating media events, such as demonstrations and surveys; debating on TV and radio; sending news releases about new developments in the campaign; and trying to get free TV and radio exposure through the Fairness Doctrine of the Federal Communications Commission.
- *Conserve limited resources* by concentrating on undecided voters, not committed opponents.
- *Get endorsements* from influential people to demonstrate a wide base of support and to encourage other community leaders to join in.
- *Know there will be disagreement* over campaign execution, so have patience and tolerance.
- *Build up to a strong campaign* at the end so the greatest exposure occurs closest to the election day.
- *Campaign door to door* to explain the legislation and provide residents with written information.
- *Place newspaper ads* toward the end of the campaign.
- *Urge sympathetic groups* to place ads in their own name rather than to simply donate money.

Gary M. Holsopple of the American Lung Association of Colorado, in Denver, points out some additional lessons from the Fort Collins campaign:

- The health issue was not the primary issue people were concerned about. Voters were concerned more about the rights of smokers and nonsmokers and the right of business people to set policy. These were the angles the Tobacco Institute used.

● The primary argument that seemed to sway the voting public in favor of Ordinance 23 was that an outsider, the tobacco industry, was telling it what is best.

● Coloradoans for Clean Indoor Air, the local citizens group, was able to address all the issues—rights, interference, health—to supplement the emphasis of other organizations, such as the Lung Association, which emphasized the health hazards of secondhand smoke.

● A brochure mailed to all registered voters advising them of the health hazards of secondhand smoke and the purpose of the ordinance helped inform all residents. This was done by the Lung Association.

● Solid support from the city council and mayor reaffirmed the right of the local community, not an outside industry, to determine policy.

## Compromise?

Lack of support by elected officials makes legislation difficult, if not impossible. Washington State FANS (Fresh Air for Non-smokers) found many legislators willing to sponsor the 1985 Washington Clean Indoor Air Act, which included workplace restrictions when necessary. Although FANS expected opposition, it didn't expect the bill to be thwarted by its own sponsors. Twelve of the co-sponsors voted with the opposition for an amendment to delete the workplace section. Jesse Wineberry, one representative who did so, defended his action this way:

> My choice was clear: to vote for progressive, though maybe not ideal, legislation. . . . A key reason why HB 62 did not have strong support was the fact that it provided no incentives for nonsmokers to negotiate. It stated that in cases

where no agreement could be reached, smoking would be prohibited. Although you may feel nonsmokers should not have to negotiate, compromise is a necessary part of all legislation.

Just what should nonsmokers negotiate about? The amount of tobacco smoke they are forced to breathe? A major point of this type of legislation is to protect nonsmokers from all tobacco smoke, just as a good organizational policy would do. The ordinance approved by San Francisco voters has a similar prohibitive clause—and the media have consistently reported that it has been working well.

Like Wineberry, many representatives who opposed the workplace section did vote for smoking restrictions in public places. They reasoned that restrictions are needed to protect the public from secondhand smoke, but workplace restrictions are government intrusion. Apparently, lobbying by the tobacco industry and the Association of Washington Business persuaded legislators that secondhand smoke is less hazardous in the workplace than elsewhere.

Deleting the workplace section wasn't enough for the tobacco industry and its supporters. They even tried (unsuccessfully) to add an amendment forbidding other communities from passing legislation stronger than the proposed state law. A related argument to squash further workplace legislation was that some of the largest employers were accommodating nonsmokers voluntarily. This argument may appeal to some lawmakers, but why should the health of employees be jeopardized because they don't work for large or progressive employers?

## Handling the "Anti-smoker" Ploy

One of the tobacco industry's favorite tactics is to cry out against "anti-smokers," people whom it depicts as militant, unreasonable, and impolite. To help counter this tactic, Bob Fox, a FANS founder, advises:

> Don't call yourself an anti-smoke group. Do not use the words No, Non, or Anti. Do use the words Clean Indoor Air, Smoke-free, and Smoke Area. By this use of words you've eliminated 90 percent of the tobacco industry's fight. They would not dare fight against clean air.

An example of the anti-smoker ploy is the article "Anti-Smoker Zealots Create Climate of Fear" in the February 1985 issue of the Tobacco Institute's bimonthly newspaper, *The Tobacco Observer*. The article tells of a woman smoking a cigarette who was reportedly harassed and barraged with ethnic slurs while waiting for a bus in San Francisco. Reports of this type, which can tarnish the campaign of nonsmokers, are exploited to the fullest by the tobacco industry. Simon Chapman discussed this problem in *The Lung Goodbye*, an organizing manual distributed by the Australian Consumers' Association:

> The smoking control lobby still labours under the unfortunate public image legacy of some of its early activists, many of whom were the last word in puritanism and everything that represents dullness. To such people, smoking was self-indulgent evil and tobacco the devil's weed. Smoking was a symptom of some more fundamental moral turpitude and so it was smokers more than smoking that was at the heart of what they reviled . . . The industry actively encourages this view and emphasizes that smoking control advocates are enemies of freedom and pleasure-haters. To diffuse this widespread preconception, it is crucial to select emphases and spokespeople who belie such images. People who carry implicit repudiation of a puritan image by their reputation (e.g: celebrities renowned for some wild, risque, or widely admired lifestyle), by their appearance or manner and by their arguments should be pushed to the front of your efforts whenever the goal is to widen community support.[2]

Chapman encourages use of the media to counter tobacco tactics—but not always in conservative ways. An outstanding example of this approach was sparked by a Philip Morris tobacco company competition for the Australian Marlboro Man. BUGA UP (Billboard Utilizing Graffitists Against Unhealthy Promotions) responded with a poster depicting a man smoking through his tracheotomy (a hole in his neck). When Philip Morris announced a winner, BUGA UP's public entry was so novel that three out of four newspaper columns were spent describing it. Chapman suggests that news conferences be held next to cancer wards and similar places to reinforce the message that smoking is hazardous to health.

## Persistence is Needed

State and local governments have been the focus of legislation for many nonsmokers' rights groups, but these decision-making bodies are not the only ones to receive attention. Some workforces, such as teachers, have their own governing boards with authority to respond to public pressure. Californians for Nonsmokers' Rights has assisted local school boards to establish smoke-free work policies for teachers, staff, and students. Likewise, well-orchestrated public pressure can force courts to provide smoking areas separate from jury deliberation rooms.

Pressure, of course, is a necessary ingredient. According to Fox: "Nothing is more important than clean indoor air where most of us spend the greatest part of our lives. You must speak out for clean air, and keep the pressure on all the time. No one is going to do it for you."

# Notes

1. *The Unabashed Self-Promoter's Guide: What Every Man, Woman, Child, and Organization in America Needs to Know About Getting Ahead and Exploiting the Media*, by Jeffrey Lant, provides incisive examples of virtually every strategy for getting publicity. It is available for $31.50 from Jeffrey Lant Associates, 50 Follen St., Suite 507, Cambridge, MA 02138.

2. *The Lung Goodbye*, by Simon Chapman (Consumer Interpol, Panang, Malaysia, 1983), originally written as a contribution to the Fifth World Conference on Smoking and Health, contains many ideas for thwarting tobacco industry promotions. It is available for $8.50 in bank draft payable to Australian Consumers' Association, 57 Carrington Road, Marrickville, New South Wales, 2204, Australia.

# Appendix 1

# Employee Attitude Survey

Employers may find a survey of their employees' attitudes helpful in implementing a policy restricting smoking. The following is a basic survey that can be modified for particular needs.

Do you smoke? Yes _____ No _____

Does tobacco smoke at work bother you? Yes _____ No _____

If it does, how so? Check all that apply.

    _____ Irritated eyes    _____ Headache

    _____ Irritated nose    _____ Nausea

    _____ Irritated throat    _____ Offensive odor

    _____ Irritated lungs    _____ Fear of consequences

    _____ Swollen sinuses

    _____ Other (specify): _____

Would you like smoking restricted at work? Yes _____ No _____

If yes, what kind of policy would you prefer?

    _____ Total ban

    _____ Designated smoking areas

Does smoke impair your work? Yes _____ No _____

If yes, how? _____

_____

How would a smoke-free workplace affect your efficiency?

    _____ Help me

    _____ Hinder me

    _____ Have no effect

How would a smoke-free workplace affect your interactions with co-workers?

    _____ It would help.

    _____ It would hinder.

    _____ It would have no effect.

If smoking restrictions were introduced, what would be the best way?

    _____ Quickly

    _____ Gradually

Do you have any additional comments?

# Appendix 2

# Checklist for Transition to a Smoke-free Environment

Will unions be involved?

Who are their representatives?

How should unions be incorporated into developing a smoke-free workplace?

Does any formal or informal policy exist already?

If an employee committee will develop policy, will there be enough members for subcommittees to do certain tasks?

If an employee committee is established, will it represent smokers, ex-smokers, nonsmokers, and unions?

Are you concerned about quit rates among smokers?

Which smokers are most important to the organization?

Might they quit? If so, who can replace them?

Are you concerned about employee attitudes toward the policy?

Are you concerned about how the policy is implemented?

Who will survey and write the evaluation several months after the policy takes affect?

What methods of policy development and implementation will ensure employee involvement?

How will employees be notified of smoke-free policy development—through memoranda, newsletters, meetings?

Should the transition to a smoke-free workplace be gradual or immediate?

Is your building sealed and without windows that open?

What kind of ventilation system does your building have?

If smoking is restricted to designated smoking areas, will a separate ventilation system be required to prevent the smoke from entering the workplace?

If smoking is restricted to designated smoking areas, will negative air pressure be needed in these areas to prevent the smoke from entering the workplace?

If you share ventilation with other firms, will you have to take into account smoke from their smokers?

Will seasonal changes be important to consider if smokers are permitted to smoke only outside?

Will the beginning date of smoking restrictions conflict with other pressures, such as heavy workloads?

How will no-smoking signs be obtained?

Who will put up the signs?

Will job application forms have to be changed?

Will job advertisements say the organization is smoke-free?

Will interviewers advise applicants of the smoke-free status?

How will you inform smoking customers of the smoke-free policy?

What kind of support will be offered to help smokers quit?

    A. Cessation programs?
       1. In-house?
       2. During working hours?
       3. For spouses?
       4. At employer's expense, or shared with smoker?
    B. Peer support system?
       1. In-house?
       2. After working hours?

What incentives will there be for the nonsmoking lifestyle?

    A. A wage increase for nonsmokers?
    B. Discretionary time off?

Additional support for a healthy lifestyle?

    A. Educational classes?
    B. Printed information?
    C. Athletic programs?
    D. Nutrition counseling?

Will you plan for a smoke-free celebration at the beginning of the policy?

    A. Will it be an open-house affair?

    B. If you invite the media, will you send information about why and how you went smoke-free?

    C. If you invite the media, will you have a spokesperson available?

# Appendix 3

# Smoke-free Consciousness-raising Checklist

1. Do you see cigarette butts strewn about?
2. Has any attempt been made to stop such littering?
3. Do smokers tend to use the restrooms to extend smoking breaks?
4. Is there evidence that cigarettes have damaged restroom floors, urinals, walls, or countertops?
5. Is there tension or controversy because of smoking in areas where smokers and nonsmokers must work together?
6. Do smokers drop their ashes on the floor whether or not ash trays are provided?
7. Are any employees absent or drawing any kind of compensation because of smoking-related illnesses?
8. Do nonsmokers place no-smoking signs in their work areas or use other means to protest unrestricted smoking?
9. Have you adapted to smokers by:
   a. Providing ashtrays?
   b. Overlooking smoker damage as "normal maintenance"?
   c. Spending money on better ventilation systems?
   d. Suppressing complaining nonsmokers?
   e. Giving in to smoker aggression?
   f. Considering smoker aggression as normal behavior?
   g. Ignoring harassment of nonsmokers from smokers and managers?
10. Have you had to extinguish a fire caused by a carelessly handled cigarette?
11. Do you consider that smoking is a right and must be allowed in the workplace?

12. Do you sympathize or side with smokers more than you do with nonsmokers over controversies about smoking at the workplace?

13. Do your carpets have cigarette burns?

14. Is disobeying a no-smoking sign an issue for disciplinary action?

15. Is it virtually impossible to police smokers even in hazardous areas where no-smoking signs are posted?

16. Do public areas such as lunchrooms have nonsmoking areas?

17. Is it considered an insult to smokers' rights to have no-smoking sections?

18. If you do have no-smoking areas, are they respected by smokers? If so, is there sufficient separation or ventilation so that drifting smoke is not a problem?

19. Have you considered that nonsmokers may be very upset about drifting smoke even though they haven't complained?

20. Have you ever reprimanded a nonsmoker for protesting about smoky conditions?

21. Are you aware that nonsmokers sometimes get upset enough about smoke to seek employment elsewhere or file a lawsuit?

22. Are nonsmokers reprimanded for extending their breaks while smokers who do the same are excused merely because they are smoking?

23. Are you aware that smokers commonly work more slowly and are absent because of illness more often than nonsmokers?

24. Are you aware that cleaning costs and insurance premiums may be reduced for companies that are smoke-free?

# Appendix 4

# Employee Morale Checklist

This checklist—compiled from the observations of executives, managers, and other employees—can help an organization see for itself how smoke affects morale:

Are nonsmokers troubled by burning eyes?

How much do nonsmokers cough or blow their nose?

Do nonsmokers put up no-smoking signs?

Do nonsmokers avoid associating with smoking co-workers, on or off the job?

Do nonsmokers post or circulate information about second-hand smoke?

What percentage of employee complaints are about smoke?

Have the number of complaints about smoke been increasing?

How much time do managers spend dealing with issues related to smoking?

Do nonsmokers move away when smokers light up?

Do nonsmokers try to wave or blow smoke away?

Do smokers and nonsmokers interact casually or tersely?

Do smokers and nonsmokers insult each other about their positions?

Are nonsmokers harassed by co-workers?

Are smokers harassed by co-workers?

Do smokers express dismay about their smoking?

Do nonsmokers look tired during breaks or after meetings?

At meetings, do nonsmokers seek seats away from smokers?

What kind of comments do nonsmokers make about the atmosphere at meetings?

Do nonsmokers look eager to leave meetings?

Do nonsmokers ask that smoking be banned at meetings?

# Appendix 5

# Calculating the Organization's Savings

Each organization should prove to itself the benefits of restricting or prohibiting smoking on the job by charting specific changes over several years. The following formulas suggest ratios that can be compared each year. A good way to do this is to calculate them over a period of five years: two years before the restrictions begin, the year they are implemented, and the two years after that.

● *Sick leave per employee per year*: Divide the total of days or hours of sick leave taken by all employees during one year by the average number of employees on the payroll that year.

● *Health insurance claims per employee per year*: Divide the number of claims taken by all employees during one year by average number of employees on payroll that year.

● *Total complaints per year*: Divide the number of formal complaints (grievances) of all types by the average number of employees on the payroll that year.

● *Smoking-related complaints per year*: Divide the number of such complaints by the average number of employees on the payroll during the same year.

● *Job-related accidents per year*: Divide the number of such accidents by the average number of employees on the payroll during the same year.

● *Resignations per year*: Divide the number of resignations submitted by the average number of employees on the payroll that year.

● *Involuntary terminations per year*: Divide the number of employees fired by the average number of employees on the payroll during the same year.

● *Early retirements per year*: Divide the number of employees who retire early by the average number on the payroll during the same year.

● *Total employee turnover*: Divide the total of employees who leave the company by the average number employed during the same year.

● *Annual maintenance costs*: Divide the costs charged to routine maintenance and housekeeping by total overhead costs.

● *Cost of disability per year*: Divide the total of disability benefits paid by the average number of employees on the payroll during the same year.

Organizations using a management control system that measures standard-cost variances can track those variances (e.g., direct labor, material, overhead) as percentages of total costs for the years preceding and following implementation of the no-smoking policy:

● *Labor efficiency*: Divide Direct Labor Efficiency Variance by total Direct Labor Cost.

● *Materials usage*: Divide Direct Material Usage Variance by total Direct Material Cost.

● *Fixed overhead*: Divide Fixed Overhead Spending Variance by total Fixed Overhead.

● *Variable overhead*: Divide Variable Overhead Price Variance by total Variable Overhead.

# Appendix 6

# Obtaining Help from Consultants

This book may contain enough information to enable you to design and implement smoking restrictions without outside help. But organizations that lack the necessary confidence to proceed may find a consultant valuable.

Because smoke-free policies are a relatively new phenomenon, the number of consultants with experience in this area is not large. Indeed, the best informed are probably activists in the nonsmokers' rights movement whose knowledge has been acquired through passionate involvement.

## Choosing a Consultant

One good way to locate a consultant is through referral from an organization listed in Appendix 10 of this book. Professional services are also advertised in business journals and newspapers. Prospects can be contacted to learn of their credentials, experience, and fees.

A good consultant should be analytical, creative, and well organized, and able to communicate clearly, both orally and in writing. Any consultant for smoke-free work policies should have expert knowledge of: (1) the scientific literature on the hazards of passive smoking; (2) the experiences of other organizations that have become smoke-free; (3) how smoking restrictions can reduce absenteeism, increase productivity, and lower the cost of maintenance and various types of insurance; (4) legal issues; and (5) how to develop a sensible plan with the fewest problems.

## Before the Consultant Arrives

Whether preparing a Formal Request for Proposals (FRP) or for an informal conference with a consultant, it would be helpful to gather the following information:

Is your goal to restrict or eliminate smoke in the workplace?

What steps have been taken so far?

What response has there been?

Are any employees represented by unions?

What problems, if any, do you expect?

What is the budget for the consultants?

What will the consultants do? Act as leaders or resource people? Make presentations to groups of employees? Attend how many meetings per week?

What resources and services can the organization provide to the consultant for support? Space, secretarial, copying, etc.?

A conference with consultants before the proposals are due is often helpful when large organizations are involved. The consultants may want information not provided in the FRP in order to develop their proposed plan.

## Management Must Be Committed

Consultants may be instrumental in developing a smoke-free workplace, but they cannot assume direct managerial roles. For effective consultation to take place, employees must know that management will support the consultant's work. If that is not the case, employees not keen on a smoke-free workplace may take advantage of the management's lack of commitment and be uncooperative with consultants. Visible executive support will go a long way toward ensuring a smooth transition.

# Appendix 7

# Summary of Legal Decisions About Smoke in the Workplace

*Shimp vs. New Jersey Bell Telephone Co.*, 368 A. 2d 408, 145 N.J. Super. 516 (1976). An office worker was granted an injunction to ensure a smoke-free area in her workplace.

*Hubbs vs. Davidson, et al.*, Mass. Super. Ct. Eq. No. 41971 (1980). A CETA trainee was granted an injunction against administrators of a job-training program, and against two trainees who were smoking, to stop smoking during the training sessions.

*Flaniken vs. Office of Personnel Management*, U.S. Merit Systems Protection Board (Dallas Field Office) 24 ATLA L. Rep. 403 (1980). An accounting technician with the Internal Revenue Service, whose severe allergy to tobacco smoke at work disabled her from performing her duties, was entitled to a disability retirement annuity, according to an administrative tribunal.

*Alexander vs. California Unemployment Insurance Appeals Board*, 163 Cal. Rptr. 411, 104 Cal. App. 3d 97 (1980). An X-ray technologist who had quit work because she was allergic to cigarette smoke and because her employer would not enforce a no-smoking policy was entitled to unemployment insurance benefits.

*Schober vs. Mountain Bell Telephone*, 96 N.M. 376, 630 P. 2d 1231 (1980). A worker who, following his collapse at work from an allergic reaction to tobacco smoke, was unable to obtain a job that would utilize his electronic skills because of his allergic reaction to tobacco smoke was disabled for purposes of workman's compensation.

*Anderson vs. Anoka County Welfare Board*, U.S. District Court

The information in this Appendix was prepared by Edward L. Sweda, Jr., Esq., for GASP of Massachusetts and appears with the organization's kind permission.

169

(D. Minn.) No. 79-269 (1981). A jury awarded $4,500 in compensatory and punitive damages for the retaliatory termination of a county social worker who complained to her department head about smoke in the office.

*Vickers vs. Veterans Administration, et al.*, 549 F. Supp 85 (1982). In a memorandum decision, the Court ruled that a V.A. worker who is "unusually sensitive to tobacco smoke" is a "handicapped person" as defined in the Federal Rehabilitation Act of 1973, but his employer had not "violated a clearly-established statutory or constitutional right" of the worker by not banning smoking in his work area.

**Smith vs. Western Electric Co.**, Missouri Court of Appeals, 643 S.W. 2d 10 (1982). A worker sought an injunction to prevent his employer from exposing him to tobacco smoke in the workplace. The trial court dismissed his petition. The Court of Appeals reversed the dismissal, saying that if Mr. Smith proves his allegations at trial, then "by failing to exercise its control and assume its responsibility to eliminate the hazardous condition caused by tobacco smoke, defendant has breached and is breaching its duty to provide a reasonably safe workplace."

*FMCS Arbitration Case*, 81 K-26042 (1982). Represented by his union, the American Federation of Government Employees' Local 1923, a federal employee submitted to arbitration. Noting the extent to which tobacco smoke can drift, the arbitrator ordered the bureaus and offices at Social Security Administration headquarters to ban smoking on the entire fifth floor of its new computer center building.

*Parodi vs. Merit Systems Protection Board*, 690 F. 2d 731 (1982). The U.S. Circuit Court of Appeals for the Ninth Circuit ruled that a government worker who is hypersensitive to smoke is "environmentally disabled" and thus eligible for disability benefits if forced to work in a smoke-filled environment. Her employer was ordered to either provide her with a smoke-free work environment or pay her disability benefits.

*Hentzel vs. Singer Co. et al.*, 188 Cal. Rptr. 159 (App. 1982). A patent attorney sued his former employer for firing him after he complained about the smoke in his office. The trial court dismissed his petition, which included a count for infliction of emotional distress. The Court of Appeals reversed the dismissal, saying that "an employee is protected against discharge or discrimination for complaining in good faith about working conditions or practices which he reasonably believes to be unsafe."

*Lee vs. Dept. of Public Welfare et al.*, Bristol County (Mass.) Superior Court No. 15385 (1983). An office worker was granted a temporary restraining order against smoking in an open office where she works with approximately 39 others, 15 of whom smoke. Request for preliminary injunction was denied. A superior court judge denied a motion to dismiss the case, saying that "an employer has no duty to make the workplace safe if, and only if, the risks at issue are inherent in the work to be done. Otherwise, the employer is required to take steps to prevent injury that are reasonable and appropriate under the circumstances. . . . Accordingly, this court cannot say that plaintiff's claim fails to make out a legally cognizable basis for relief." In an out-of-court settlement, the state agreed to provide nonsmokers with smoke-free offices and other important safeguards.

# Appendix 8

# Organizations With Smoking Restrictions

The following are examples of organizations that recognize workers' needs for clean air.

## Some Smoke-free Areas

Aetna Life and Casualty Co., Hartford, Conn.
Chicago Tribune, Chicago, Ill.
Cincinnati Bell Telephone Co., Cincinnati, Ohio
Continental Illinois Bank and Trust Co., Chicago, Ill.
E. E. du Pont de Nemours and Co., Wilmington, Del., and elsewhere
Frederick Electronics, Frederick, Md.
Health and Welfare, Canada
J. C. Penney, Inc., New York, N.Y.
Levi-Strauss, San Francisco, Calif.
Martin Marietta Corporation, Bethesda, Md., and elsewhere
McGraw-Hill, New York, N.Y.
Mobil Oil Corporation, New York, N.Y.
New England Mutual Life Insurance Co., Boston, Mass.
Sperry Co., Blue Bell, Pa.

---

From *Toward A Smoke-Free Workplace*, by Regina Carlson [New Jersey Group Against Smoking Pollution, Inc., 1985], an excellent manual available for $5.00 from New Jersey GASP, 105 Mountain Ave., Summit, NJ 07901.

## More Extensive Smoke-free Areas, Including Work Stations

Bancroft and Whitney Co., San Francisco, Calif.
British Columbia Ministry of Health, Canada
Brooklyn, N.Y., District Attorney
CIGNA Insurance, Philadelphia, Pa.
Citizen, Northshore, Wash.
City Federal Savings and Loan (throughout New Jersey)
Control Data Corporation, Minneapolis, Minn.
Florida Dept. of Health and Rehabilitative Services, Tallahassee, Fla.
Forum, Hackettstown, N.J.
General Telephone Co. of the Northwest, Inc., Everett, Wash.
Grumman Corporation, New York, Florida, and California
Harvard University, Cambridge, Mass.
Honeywell, Inc., Minneapolis, Minn.
Lawrence-Berkeley Laboratories, Berkeley, Calif.
Los Angeles, California, airport air traffic controllers' worksite
Marion County Health Dept., Indianapolis, Ind.
Midland Brake Inc., Iola, Kans.
W. W. Richardson Insurance Agency, Inc., Warren, R.I.
Wall Street Journal, New York, N.Y.

## Smoke-free Except for Cafeteria Areas and Certain Lounges

Adolph Coors Co., Golden, Colo.
Adrian Construction Co., Inc., Dallas, Tex.
Becton Dickson and Co., Paramus, N.J. (except private offices)
Blue Cross-Blue Shield of Minnesota
Campbell Soup Co., Camden, N.J.
Central Telephone Company-Nevada, Las Vegas, Nev.

Falcon Safety Products, Mountainside, N.J.
Family Life Insurance, Seattle, Wash.
Federal Cartridge Corporation, Anoka, Minn.
Fusion Systems Corporation, Rockville, Md.
Juneau-Douglas Telephone Co., Juneau, Alaska
Lee Co., Salina, Kans.
Medtronic, Minneapolis, Minn.
New Brunswick Scientific Co., Inc., Edison, N.J.
New England Deaconess Hospital, Boston, Mass.
Premier Dental Products Co., Norristown, Pa.
Raven Industries, Sioux Falls, S.D.
Westlake Hospital, Melrose Park, Ill.

## Entirely Smoke-free

The Aerobics Center, Dallas, Tex.
American Biltrite, Trenton, N.J.
American Heart Association National Center, Dallas, Tex.
American Lung Association, nationwide
Andover, Kans., schools
Benton, Ark., schools
British Columbia Hydro and Power Authority, Vancouver, B.C.
Fargo Electronics, Eden Prairie, Minn.
Group Health Cooperative of Puget Sound, Seattle, Wash.
Hinsdale Hospital, Hinsdale, Ill.
Joan Eastman Bennett Property Designs, Summit, N.J.
Logo Computer, Boston, Mass.
Moselle Insurance Inc., Los Angeles, Calif.
MPD Printing, Summit, N.J.
Nesley Allen Brass Beds, Los Angeles, Calif.
Pima Care, Inc., Tucson, Ariz.
Rodale Press, Emmaus, Pa.
Salina, Kans., Journal

Seaborg, Inc., Huntington Beach, Calif.
Slack, Inc., West Deptford, N.J.
Spenco Medical Corporation, Waco, Tex.
Stride-Rite, Cambridge, Mass.
Surrogate Hostess Restaurant, Seattle, Wash. (including patrons)
Tip Top Printing Co., Daytona Beach, Fla.
Town Crier, Wayland-Sudbury, Mass.
Vanguard Electronic Tool Co., Seattle, Wash.
WRNJ, Hackettstown, N.J.
Zycad, Inc., Arden Hills, Minn.

## Employers that Hire Only Nonsmokers

Adrian Construction Co., Inc., Dallas, Tex. (preference to non-
smokers)
Alexandria, Virginia, Fire Department
Black Hills Hospital, Olympia, Wash.
Dean Equipment and Furniture Co., Inc., Fairfield, N.J.
Fortunoff's, New York, N.Y.
Quin-T Corporation, Tilton, N.J.
Salem, Ore., Fire Department
Wayne Green Enterprises, Peterborough, N.H.
Westminster Office Machines, Inc., Lake Bluff, Ill.
Wichita, Kans., Fire Department

# Appendix 9

# Model Smoking
# Pollution Control Ordinance

Sec. 1000    Title    This Article shall be known as the Smoking Pollution Control Ordinance.

Sec. 1001    Purpose    WHEREAS, numerous studies have found that tobacco smoke is a major contributor to indoor air pollution; and

---

This ordinance, developed by Californians for Nonsmokers' Rights, has been used as the basis for laws in California and elsewhere. Revised editions, issued periodically, will be available for $5.00 from Californians for Nonsmokers' Rights, 2054 University Avenue, Suite 500, Berkeley, CA 94704.

WHEREAS, studies have shown involuntary smoking to be a significant health hazard for several populations, including elderly people, individuals with cardiovascular disease, and individuals with impaired respiratory function, including asthmatics and those with obstructive airway disease; and

WHEREAS, health hazards induced by involuntary smoking include lung cancer, respiratory infection, decreased exercise tolerance, decreased respiratory function, bronchoconstriction, and bronchospasm; and

WHEREAS, nonsmokers with allergies, respiratory diseases and those who suffer other ill effects of breathing secondhand smoke may experience a loss of job productivity or may be forced to take periodic sick leave because of reactions to secondhand smoke; and

WHEREAS, studies have shown many nonsmokers do not dine in restaurants because of adverse reaction or annoyance from secondhand smoke.

WHEREAS, the vast majority of travelers prefer nonsmoking sections in airplanes, buses, and trains;

WHEREAS, smoking is a potential cause of fires, and cigarette and cigar burns and ash stains on merchandise and fixtures cause losses to Sacramento businesses;

NOW, therefore, the Board of Supervisors (City Council) finds and declares that the purposes of this ordinance are (1) to protect the public health and welfare by prohibiting smoking in public places and places of employment except in designated smoking areas and (2) to strike a reasonable balance between the needs of persons who smoke and the need of nonsmokers to breathe smoke-free air, and to recognize that, where these needs conflict, the need to breathe smoke-free air shall have priority.

Sec. 1002    Definitions    The following words and phrases, whenever used in this ordinance, shall be construed as defined in this section:

1. "Bar" means an area which is devoted to the serving of alcoholic beverages for consumption by guests on the premises and in which the serving of food is only incidental to the consumption of such beverages. Although a restaurant may contain a bar, the term "bar" shall not include the restaurant dining area.

2. "Business" means any sole proprietorship, partnership, joint venture, corporation or other business entity formed for profit-making purposes, including retail establishments where goods or services are sold as well as professional corporations and other entities under which legal, medical, dental, engineering, architectural or other professional services are delivered.

3. "Dining Area" means any enclosed area containing a counter or tables upon which meals are served.

4. "Employee" means any person who is employed by any employer in the consideration for direct or indirect monetary wages or profit.

5. "Employer" means any person who employs the services of an individual person.

6. "Enclosed" means closed in by a roof and four (4) walls with appropriate openings for ingress and egress.

7. "Motion Picture Theater" means any theater engaged in the business of exhibiting motion pictures.

8. "Non-Profit Entity" means any corporation, unincorporated association or other entity created for charitable, philanthropic, educational, character building, political, social or other similar purposes, the net proceeds from the operations of which are committed to the promotion of the objects or purposes of the organization and not to

private financial gain. A public agency is not a "Non-Profit Entity" within the meaning of this Section.

9. "Place of Employment" means any enclosed area under the control of a public or private employer which employees normally frequent during the course of employment, including, but not limited to,

    1. conference and class rooms
    2. employee cafeterias
    3. employee lounges and restrooms
    4. hallways
    5. work areas
    6. A private residence is not a "place of employment" unless said residence is used as a child care or a health care facility.
    7. The dining area of a restaurant is not a "place of employment."
    8. It is not our intent to regulate governmental agencies not under the jurisdiction of the city or county.
    9. The intent is to include nonprofit corporations, offices and other facilities maintained by pubic agencies which are under the jurisdiction of the city or county, and other entities not commonly understood to be business enterprises, though frequented by the public.

10. "Public Place" means any enclosed area to which the public is invited or in which the public is permitted, including, but not limited to:

    1. banks
    2. educational facilities
    3. health facilities
    4. public transportation facilities

    5. reception areas
    6. restaurants
    7. retail stores
    8. retail service establishments
    9. retail food production and marketing establishments
   10. waiting rooms
   11. A private residence is not a "public place."

11. "Restaurant" means any coffee shop, cafeteria, luncheonette, tavern, cocktail lounge, sandwich stand, soda fountain, private and public school cafeteria or eating establishment, and any other eating establishment, organization, club, including veterans' club, boardinghouse, or guesthouse which gives or offers for sale food to the public, guests, patrons, or employees as well as kitchens in which food is prepared on the premises for serving elsewhere, including catering functions, except that the term "restaurant" shall not include a cocktail lounge or tavern if said cocktail lounge or tavern is a "bar" as defined in section 1002.1.

12. "Service Line" means any indoor line at which one (1) or more persons are waiting for or receiving service of any kind, whether or not such service involves the exchange of money.

13. "Smoking" means inhaling, exhaling, burning or carrying any lighted cigar, cigarette, weed, plant or other combustible substance in any manner or in any form.

14. "Sports Arena" means sports pavilions, gynmasiums, health spas, boxing arenas, swimming pools, roller and ice rinks, bowling alleys and other similar places where members of the general public assemble to either engage in physical exercise, participate in ath ᶜ competition or witness sports events.

15. "Tobacco Store" means a retail store utilized primarily for the sale of tobacco products and accessories and in which the sale of other products is merely incidental.
16. "Work Area" or "Workplace" means any area of a place of employment enclosed by floor to ceiling walls in which two or more employees are assigned to perform work for an employer.

Sec. 1003     Smoking Regulated—Places of Employment     It shall be the responsibility of employers to provide smoke-free areas for nonsmokers within existing facilities to the maximum extent possible, but employers are not required to incur any expense to make structural or other physical modifications in providing these areas.

Within 90 days of the effective date of this ordinance, each employer shall adopt, implement, make known and maintain a written smoking policy which shall contain as a minimum the following requirements:

1. Any employee in a place of employment shall be given the right to designate his or her work area as a non-smoking area and to post the same with an appropriate sign or signs.
2. Provision and maintenance of a separate and contiguous nonsmoking area of not less then fifty percent (50%) of the seating capacity and floor space in cafeterias, lunchrooms an employee lounges or provision and maintenance of separate and equal sized cafeterias, lunchrooms and employee lounges for smokers and nonsmokers.
3. Prohibition of smoking in employer:

    1. auditoriums
    2. classrooms
    3. conference and meeting rooms

4. elevators
5. hallways (some employers may want to permit smoking in hallways and bathrooms as a way to prohibit smoking in the remainder of the work area)
6. medical facilities
7. restrooms.

4. In any dispute arising under the smoking policy, the rights of the nonsmoker shall be given precedence.
5. The smoking policy shall be communicated to all employees within three (3) weeks of its adoption, and at least yearly thereafter.
6. All employers shall supply a written copy of the smoking policy to any prospective employee who so requests.
7. Notwithstanding the provisions of this section, every employer shall have the right to designate any place of employment, or any portion thereof, as a nonsmoking area.
8. An employer who makes reasonable efforts to develop and promulgate a policy regarding smoking and and nonsmoking in the workplace shall be deemed to be in compliance with this section, provided that a policy which knowingly omits the minimal elements of a policy or which designates an entire workplace as a smoking area shall not be deemed in compliance with this section.
9. A private enclosed office workplace occupied exclusively by smokers, even though such an office workplace may be visited by nonsmokers, may be declared a smoking area.

Sec. 1004    Smoking Optional    Notwithstanding any other provision of this article to the contrary, the following areas shall not be subject to the smoking restrictions of the article:

1. Bars
2. Private residences, except when used as a child care or health care facility

3. Hotel and motel rooms rented to guests
4. Retail tobacco stores
5. Eating establishments, hotel and motel conference/meeting rooms, and public and private assembly rooms while these places are being used for private functions
6. A private residence which may serve as a place of employment
7. A private enclosed office workplace occupied exclusively by smokers, even though such an office workplace may be visited by nonsmokers
8. Semiprivate rooms of health facilities occupied by one (1) or more patients, all of whom are smokers who have requested in writing on the health care facilities' admissions forms to be placed in a room where smoking is permitted.
9. Notwithstanding any other provision of this section, any owner, operator, manager or other person who controls a business may declare that entire business as a nonsmoking establishment.

Sec. 1005    Smoking Prohibited    Smoking shall be prohibited in the following places within XYZ County:

1. Elevators.
2. Buses, taxicabs, and other means of public transit under the authority of Sacramento County, while within the boundaries of the county, and in ticket, boarding, and waiting areas of public transit depots; provided, however, this prohibition does not prevent the establishment of separate equal sized waiting areas for smokers and nonsmokers, or establishing no more than fifty percent (50%) of a given waiting area as a smoking area.
3. Public restrooms.
4. Indoor service lines.

5. Retail stores doing business with the general public, except areas in said stores not open to the public and all areas within retail tobacco stores.

6. All enclosed areas available to and customarily used by the general public in all businesses or non-profit entities patronized by the public, including, but not limited to, attorney's offices and other offices, banks, hotels and motels. The intent is to include non-profit corporations, offices and other facilities maintained by public agencies which are under the jurisdiction of the city or county, and other entities not commonly understood to be business enterprises, though frequented by the public.

7. Within all restaurants and other eating establishments, provided, however, this prohibition does not prevent the designating of a contiguous area within the restaurant that contains no more than fifty percent (50%) of the seating capacity of the restaurant and a smoking area or the providing of separate rooms for smokers and non-smokers provided the rooms designated as smoking do not contain more than fifty percent (50%) of the seating capacity of the restaurant.

8. In public areas of aquariums, libraries, and museums when open to the public, provided however, this prohibition does not prevent the designation of a separate room for smoking.

9. Within any building not open to the sky which is primarily used for, or designed for the primary purpose of exhibiting any motion picture, stage drama, lecture, musical recital or other similar performance, except when smoking is part of a stage production; provided however, this prohibition does not prevent designating a contiguous area containing no more than fifty percent (50%) of any any commonly called a lobby as a smoking area.

10. Within sports arenas and convention halls, except in designated smoking areas.

11. Within every room, chamber, place of meeting or public assembly, including school buildings under the control of any board, council, commission, committee, including joint committees, or agencies of the County (or City) or any political subdivision of the State during such time as a public meeting is in progress.
12. Waiting rooms, hallways, wards and semiprivate rooms of health facilities, including but not limited to, hospitals, clinics, physical therapy facilities, doctors' and dentists' offices, except in separate designated smoking areas.
13. In bed space areas utilized for two or more patients, smoking shall be prohibited unless all patients within the room are smokers and request in writing upon the health care facility's admission forms to be placed in a room where smoking is permitted.
14. Notwithstanding any other provision of this section, any owner, operator, manager or other person who controls any establishment described in this section may declare that entire establishment as a nonsmoking establishment.

## Sec. 1006    Posting of Signs

1. "Smoking or "No Smoking" signs, whichever are appropriate, with letters of not less than one inch (1″) in height or the international "No Smoking" symbol (consisting of a pictorial representation of a burning cigarette enclosed in a red circle with a red bar across it) shall be clearly, sufficiently and conspicuously posted in every room, building or other place where smoking is regulated by this chapter by the owner, operator, manager or other person having control of such building or other place.
2. Every theater owner, manager or operator shall post signs conspicuously in the lobby stating that smoking is prohibited within the theater or auditorium, and in the case of motion picture theaters, such information shall be

shown upon the screen for at least five (5) seconds prior to the showing of each feature motion picture.

3. Every restaurant will have posted at its entrance a sign clearly stating that a nonsmoking section is available, and every patron shall be asked as to his or her preference.

## Sec. 1007     Enforcement

1. Enforcement shall be implemented by the County Administration Officer or his or her designees (or the City Manager).

2. Any citizen who desires to register a complaint under this chapter may initiate enforcement with the County Administration Officer, or his or her designees (or the City Manager).

3. The fire department or the health department shall require while an establishment is undergoing otherwise mandated inspections, a "self-certification" from the owner, manager, operator or other person having control of such establishment that all requirements of this chapter have been complied with.

4. Any owner, manager, operator or employee of any establishment controlled by this chapter may inform persons violating this article of the appropriate provisions thereof.

5. Notwithstanding any provision of this chapter, a private citizen may bring legal action to enforce this chapter.

## Sec. 1008     Violations and Penalties

1. It shall be unlawful for any person who owns, manages, operates or otherwise controls the use of any premises subject to the restrictions of this chapter to fail to comply with its provisions.

2. The owner, manager or operator of a restaurant shall not be held liable if their host or hostess fails to ask the

seating preference of patrons. They will be held liable if the restaurant has no stated policy concerning asking the preference of patrons.

3. It shall be unlawful for any person to smoke in any area restricted by the provisions of this chapter. (Intent is to exclude employees from liability under section 1003.)

4. Any person who violates any provision of this article shall be guilty of an infraction, punishable by:

5. (1) A fine not exceeding one hundred dollars ($100) for a first violation. (2) A fine not exceeding two hundred dollars ($200) for a second violation. (3) A fine not exceeding five hundred dollars ($500) for each additional violation of this article within one (1) year.

**Sec. 1009     Nonretaliation**     No person or employer shall discharge, refuse to hire, or in any manner retaliate against any employee or applicant for employment because such employee or applicant exercises any rights afforded by this chapter.

**Sec. 1010     Severability**     If any provision, clause, sentence or paragraph of this chapter or the application thereof to any person or circumstances shall be held invalid, such invalidity shall not affect the other provisions of this chapter which can be given effect without the invalid provision or application, and to this end the provisions of this chapter are declared to be severable.

**Sec. 1011     Public Education**     The County Executive Officer (or City Manager) shall engage in a continuing program to inform and clarify the purposes of this chapter to citizens affected by it, and to guide owners, operators, and managers in their compliance.

The County Executive Officer (or City Manager) shall leave the responsibility of conducting a public education campaign, regarding the health consequences of smoking to other governmental and health agencies equipped with the needed expertise to conduct such campaigns.

Sec. 1012    Governmental Agency Cooperation    The County Executive Officer (or City Manager) shall annually request such governmental and educational agencies having offices within the County of Sacramento to establish local operating procedures to cooperate and comply with this chapter. In Federal, State and special school districts within the County of XYZ, the County Executive Officer (or City Manager) shall urge enforcement of their existing no smoking prohibitions and request cooperation with this chapter.

(It is not our intent to regulate any governmental agency not currently under the jurisdiction of XYZ County.)

Sec. 1013    Other Applicable Laws    This chapter shall not be interpreted or construed to permit smoking where it is otherwise restricted by other applicable health, safety or fire codes.

Sec. 1014    Exemptions

1. Any owner or manager of a business or other establishment subject to this chapter may apply to the county for an exemption or modification to any provision of this article due to unusual circumstances or conditions. Intent is to exempt on a showing of financial impracticability.
2. Such exemption shall be granted only if the County Executive finds from the evidence presented by the applicant for exemption at a public hearing.
3. The applicant for an exemption shall pay the fee prescribed in the zoning code for a use permit with the application to cover the cost of the hearing and noticing of the hearing.

Sec. 1015    Effective date    This chapter shall take effect within 30 days of passage, and shall be reviewed one year after its effective date.

# Appendix 10

# Nonsmokers' Rights Organizations

## United States

Action on Smoking and Health, 2013 H St. NW, Washington, DC 20006

Akron GASP, P.O. Box 8338, Akron, OH 44320

Alaska Council on Smoking or Health, P.O. Box 3-4028, Anchorage, AK 99501

American Council on Science and Health, 1995 Broadway, 18th Floor, New York, NY 10023

American Lung Association, 1740 Broadway, New York, NY 10019

Anchor Organization, 1725 W. Harrison St., Chicago, IL 60612

Arizonans Concerned About Smoking, 11801 North Sundown Lane, Scottsdale, AZ 85254

Association for Nonsmokers' Rights (ANSR), 1829 Portland Ave. S., Minneapolis, MN 55404

Bowie GASP, P.O. Box 863, Bowie, MD 20715

Californians for Nonsmokers' Rights (CNR), P.O. Box 668, Berkeley, CA 94701

Californians for Nonsmokers' Rights (CNR), Orange County Affiliate, P.O. Box 1543, Santa Ana, CA 92702

Californians for Nonsmokers' Rights (CNR), Sacramento Affiliate, 909 12th St., Sacramento, CA 95814

Californians for Nonsmokers' Rights (CNR), San Diego Affiliate, P.O. Box 99511, San Diego, CA 92109

Californians for Nonsmokers' Rights (CNR), San Fernando Valley Affiliate, P.O. Box 89, Van Nuys, CA 91406

Californians for Nonsmokers' Rights (CNR), Westside Affiliate, P.O. Box 3225, Santa Monica, CA 90403

Caring for Nonsmokers, 7022 S. Jan Mar Dr., Dallas, TX 75230

Center for Nonsmokers' Rights, 2007 Broadway, Kansas City, MO 64108

Cincinnati GASP, P.O. Box 37731, Cincinnati, OH 45222

Citizens for Clean Indoor Air, c/o Western Missouri Lung Association, 2007 Broadway, Kansas City, MO 64108

Clean Air Team, P.O. Box 4349, Honolulu, HI 96813

Cleveland GASP, P.O. Box 14871, Cleveland, OH 44114

DOC (Doctors Ought to Care), c/o Rick Richards, M.D., 3942 Willowood Road, Martinez, GA 30907

Environmental Health Committee, P.O. Box 23817, San Jose, CA 95135

Environmental Improvement Associates, 109 Chestnut St, Salem, NJ 08079

ERNS, 2502 Sombrosa Place, Carlsbad, CA 92008

FANS (Fresh Air for Nonsmokers), P.O. Box 180751, Austin, TX 78718

Fresh Air for Nonsmokers (FANS), P.O. Box 24052, Seattle, WA 98124

Fresh Air for Nonsmokers (FANS), P.O. Box 1357, Spokane, WA 99210

GASP (Group Against Smoking Pollution), P.O. Box 450981, Atlanta, GA 30345

GASP, P.O. Box 666, Bismarck, ND 58502

GASP, Rt 2, P.O. Box 665, Bonneau, SC 29431

GASP, Georgia Lung Assoc., 1383 Spring St, Atlanta, GA 30367

GASP in Tallahassee, P.O. Box 13672, Tallahassee, FL 32317

GASP (National Office), P.O. Box 632, College Park, MD 20740

GASP of Arkansas, P.O. Box 7697, Little Rock, AK 72217

GASP of Colorado, P.O. Box 12103, Boulder, CO 80303

GASP of Maine, 47 Thomas St, Portland, ME 04102

GASP of Massachusetts, 99 Brookline Ave., Boston, MA 02215

GASP of Miami, P.O. Box 45-0952, Miami, FL 33145

GASP of New York, Box 101, Plainview, NY 11804

GASP of Northern Colorado, P.O. Box 2306, Ft. Collins, CO 80522

Hames Nonsmokers' Rights Groups, 354 N. Glenn Ave, Fresno, CA 93701

Hawaiian Islands Nonsmokers Organization (HINO), P.O. Box 10-Q, Honolulu, HI 95816

Hold the Smoke, P.O. Box 1594, Aspen, CO 81612

Kansans for Nonsmokers' Rights, 1311 W. 5th St., Topeka, KS 66606

Lehigh Valley Committee Against Health Fraud, Inc., P.O. Box 1747, Allentown, PA 18105

National Council Against Health Fraud, Inc., P.O. Box 1276, Loma Linda, CA 92354

National Interagency Council on Smoking and Health, c/o American Heart Association, 7320 Greenville Ave, Dallas, TX 75231

New Jersey GASP, 105 Mountain Ave, Summit, NJ 07901

New Jersey Interagency Council on Smoking and Health, 129 E. Hanover St., 4th Floor CN 362, Trenton, NJ 08608

New Mexico Nonsmoker Protection Projects, P.O. Box 657, Los Alamos, NM 87544

Non Smokers' Rights Committee, American Lung Association of Northeast Pennsylvania, P.O. Box 115, Scranton, PA 18504

Nonsmokers Incorporated, 7 South Fluorite, Tucson, AZ 85745.

Nonsmokers Travel Club of GASP, 8929 Bradmoor Dr., Bethesda, MD 20034

Non-Smokers Unite, P.O. Box 1083, Troy, MI 48099

Northern Virginia GASP, P.O. Box 442, Springfield, VA 22150

Ohio GASP, P.O. Box 14871, Cleveland, OH 44114

Pittsburgh Association for Nonsmokers' Rights (ANSR), P.O. Box 4983, Pittsburgh, PA 15206

Please No Smoking Club, P.O. Box 25972, Albuquerque, NM 87125

Right to Breathe, P.O. Box 7772, Naples, FL 33941

St. Louis GASP, P.O. Box 6086, Southwest Station, St. Louis, MO 63139

San Gabriel-Pomona Valley GASP, P.O. Box 10155, Glendale, CA 91209

Smoke Signal, P.O. Box 99688, San Francisco, CA 94109

Spokane FANS, P.O. Box 1357, Spokane, WA 99210

Texans Against Public Smoking, P.O. Box 19847, Houston, TX 77024

Tobacco Education Council, P.O. Box 6776, Ventura, CA 93006

Wichita GASP, P.O. Box 17062, Wichita, KS 67217

## Australia

BUGA UP, P.O. Box 78, Wentworth Building, University of Sydney Union, Sydney 2006

BUGA UP, P.O. Box 285, Fitzroy, Victoria 3065

## Belgium

A.I.R. P.U.R., Michel Loisseau-T, Rue de Thier a Liege 142, B-4000 Liege

## Canada

Canadian Council on Smoking and Health, 725 Churchill Ave, Ottawa, Ontario K1Z 5G7

GASP Edmonton, 10220 Churchill Cres, Edmonton, Alberta

GASP, P.O. Box 911, Winnipeg, Manitoba

Hamilton-Wentworth Nonsmokers' Rights Association, P.O. Box 33, Stony Creek, Ontario L8G 3X7

London and Middlesex Lung Association, 148 York St., Fredericton, New Brunswick E3B 5B4

Non-Smokers' Rights Association, 455 Spadina Ave., Suite 201, Toronto, Ontario M5S 2G8

## Israel

Anti Smoking Public Information and Education, Israel Cancer Association, 91, Hachasmonaim St, Tel-Aviv
Israel Council to Prevent Smoking, P.O. Box 4131, Jerusalem

## Japan

Citizens Action for Nonsmokers' Rights, 1-7-3 Hirakawa-chiyoda-ku, Tokyo
Lawyers' Organization for Nonsmokers' Rights, c/o Yoshio Isa-yama, Yotsuya Law Office, 7th Floor, ITO Building, 1-2 Yotsuya, Shinjuku-ku, Tokyo

## Norway

Tobakken OG VI, Minister Ditleffs Vei 17c, Oslo 8

## Switzerland

Swiss Association of Non-Smokers, Kohlrainstrasse 5, CH-8700 Kusnacht

## United Kingdom

Action on Smoking and Health, 5, Mortimer St, London WIN7RH

Association for Nonsmokers' Rights, 82 Stephens St, Edinburgh
EH35AQ
National Society of Non Smokers, Information Center, Latimer
House, 4048 Hanson St., London W1
Nonsmokers' Campaign, 1 Birdbush Avenue, Saffron Walden,
Essex

## Zimbabwe

Organization Against Smoking in Society, Dept. of Surgery,
University of Zimbabwe, POB MP167, Mt. Pleasant, Harare,
Zimbabwe